Springer Series in

OPTICAL SCIENCES 89

founded by H.K.V. Lotsch

Springer
Berlin
Heidelberg
New York
Hong Kong
London
Milan
Paris
Tokyo

Springer Series in
OPTICAL SCIENCES

The Springer Series in Optical Sciences, under the leadership of Editor-in-Chief *William T. Rhodes*, Georgia Institute of Technology, USA, and Georgia Tech Lorraine, France, provides an expanding selection of research monographs in all major areas of optics: lasers and quantum optics, ultrafast phenomena, optical spectroscopy techniques, optoelectronics, quantum information, information optics, applied laser technology, industrial applications, and other topics of contemporary interest.
With this broad coverage of topics, the series is of use to all research scientists and engineers who need up-to-date reference books.

The editors encourage prospective authors to correspond with them in advance of submitting a manuscript. Submission of manuscripts should be made to the Editor-in-Chief or one of the Editors. See also http://www.springer.de/phys/books/optical_science/

Motoichi Ohtsu (Ed.)

Progress in Nano-Electro-Optics II

Novel Devices and Atom Manipulation

With 112 Figures and 3 Tables

Professor Dr. Motoichi Ohtsu
Tokyo Institute of Technology
Interdisciplinary Graduate School
of Science and Engineering
4259 Nagatsuta-cho, Midori-ku,
Yokohama 226-8502,
Japan
E-mail: ohtsu@ae.titech.ac.jp

ISSN 0342-4111

ISBN 3-540-05042-6 Springer-Verlag Berlin Heidelberg New York

Library of Congress Cataloging-in-Publication Data

Progress in nano-electro-optics II : novel devices and atom manipulation / Motoichi Ohtsu (ed.). p.cm. --
(Springer series in optical sciences ; v. 89)
Includes bibliographical references and index.
ISBN 3540050426 (alk. paper)
1. Electrooptics. 2. Nanotechnology. 3. Near-field microscopy. I. Ohtsu, Motoichi. II. Series.
TA1750 .P75 2002 621.381'045--dc21 2002030321

Springer-Verlag Berlin Heidelberg New York
a member of BertelsmannSpringer Science+Business Media GmbH

http://www.springer.de

Camera-ready by the author using a Springer TEX macropackage
Final processing and production by LE-TeX Jelonek, Schmidt & Vöckler GbR, Leipzig
Cover concept by eStudio Calamar Steinen using a background picture from The Optics Project. Courtesy of John T. Foley, Professor, Department of Physics and Astronomy, Mississippi State University, USA.
Cover production: *design & production* GmbH, Heidelberg

Printed on acid-free paper 57/3141/YL 5 4 3 2 1 0

Preface to *Progress in Nano-Electro-Optics*

Recent advances in electro-optical systems demand drastic increases in the degree of integration of photonic and electronic devices for large-capacity and ultrahigh-speed signal transmission and information processing. Device size has to be scaled down to nanometric dimensions to meet this requirement, which will become even more strict in the future. In the case of photonic devices, this requirement cannot be met only by decreasing the sizes of materials. It is indispensable to decrease the size of the electromagnetic field used as a carrier for signal transmission. Such a decrease in the size of the electromagnetic field beyond the diffraction limit of the propagating field can be realized in optical near fields.

Near-field optics has progressed rapidly in elucidating the science and technology of such fields. Exploiting an essential feature of optical near fields, i.e., the resonant interaction between electromagnetic fields and matter in nanometric regions, important applications and new directions such as studies in spatially resolved spectroscopy, nano-fabrication, nano-photonic devices, ultrahigh-density optical memory, and atom manipulation have been realized and significant progress has been reported. Since nano-technology for fabricating nanometric materials has progressed simultaneously, combining the products of these studies can open new fields to meet the above-described requirements of future technologies.

This unique monograph series entitled "Progress in Nano-Electro-Optics" is being introduced to review the results of advanced studies in the field of electro-optics at nanometric scales and covers the most recent topics of theoretical and experimental interest on relevant fields of study (e.g., classical and quantum optics, organic and inorganic material science and technology, surface science, spectroscopy, atom manipulation, photonics, and electronics). Each chapter is written by leading scientists in the relevant field. Thus, high-quality scientific and technical information is provided to scientists, engineers, and students who are and will be engaged in nano-electro-optics and nano-photonics research.

I gratefully thank the members of the editorial advisory board for valuable suggestions and comments on organizing this monograph series. I wish to express my special thanks to Dr. T. Asakura, Editor of the Springer Series in Optical Sciences, Professor Emeritus, Hokkaido University for recommending me to publish this monograph series. Finally, I extend an acknowledgement to Dr. Claus Ascheron of Springer-Verlag, for his guidance and suggestions, and to Dr. H. Ito, an associate editor, for his assistance throughout the preparation of this monograph series.

Yokohama, October 2002 *Motoichi Ohtsu*

Preface to Volume II

This volume contains four reviews chapters focusing on different aspects of nano-electro-optics. The first chapter deals with classical theory on electromagnetic near field in the vicinity of a matter. The motivation of this work is to answer the question "Why a resolution far beyond the diffraction limit is attained in near-field optical microscopy?"

The second chapter is devoted to a review of the fundamental aspects of excitonic polaritons that are propagating in quantum-well waveguides. The authors demonstrate the possible application of these excitonic polaritons to optoelectronic devices. Reduction of the optical-switch size to the nanometer scale is also discussed.

The third chapter concerns the instrumentation and measurements of near-field optical microscopy and its application to imaging spectroscopy of single-quantum constituents in order to study the intrinsic nature of quantum-confined systems. Real-space mapping of exciton wavefunctions confined in a quantum dot is also demonstrated.

The final chapter deals with atom manipulation by optical near field. The authors introduce a slit-type atom deflector and a detector, which are fabricated from a silicon-on-insulator substrate. They also claims that atom-control techniques by optical near field will greatly develop nanophotonics and create a new research field, atom-photonics.

As was the case of volume I, this volume is published by the support of an associate editor and members of an editorial advisory board. They are:

Associate editor: Ito, H. (Tokyo Inst. Tech., Japan)

Editorial advisory board: Barbara, P.F. (Univ. of Texas, USA)
Bernt, R. (Univ. of Kiel, Germany)
Courjon, D. (Univ. de Franche-Comte, France)
Hori, H. (Yamanashi Univ., Japan)
Kawata, S. (Osaka Univ., Japan)
Pohl, D. (Univ. of Basel, Switzerland)
Tsukada, M. (Univ. of Tokyo, Japan)
Zhu, X. (Peking Univ., China)

I hope that this volume will be a valuable resource for the readers and future specialists.

Yokohama, April 2003 *Motoichi Ohtsu*

Preface to Volume II

Contents

List of Contributors

Itsuki Banno
Faculty of Engineering
University of Yamanashi
Kofu, Yamanashi 400-8511, Japan
banno@es.yamanashi.ac.jp

Kazuhiko Hosomi
Nanoelectronics Collaborative
Research Center
Institute of Industrial Science
The University of Tokyo
4-6-1 Komaba, Meguro-ku
Tokyo 153-8505, Japan
hosomi@iis.u-tokyo.ac.jp

Haruhiko Ito
Interdisciplinary Graduate School
of Science and Technology
Tokyo Institute of Technology
4259 Nagatsuta-cho, Midori-ku
Yokohama 226-8502, Japan
ito@ae.titech.ac.jp

Toshio Katsuyama
Nanoelectronics Collaborative
Research Center
Institute of Industrial Science
The University of Tokyo
4-6-1 Komaba, Meguro-ku
Tokyo 153-8505, Japan
katsuyam@iis.u-tokyo.ac.jp

Motoichi Ohtsu
Interdisciplinary Graduate School
of Science and Technology
Tokyo Institute of Technology
4259 Nagatsuta-cho, Midori-ku
Yokohama 226-8502, Japan
ohtsu@ae.titech.ac.jp

Toshiharu Saiki
Department of Electronics
and Electrical Engineering
Keio University
3-14-1 Hiyoshi, Kohoku-ku
Yokohama 223-8522, Japan
saiki@elec.keio.ac.jp

Kouki Totsuka
ERATO Localized Photon Project
Japan Science and Technology
Corporation
687-1 Tsuruma
Machida, Tokyo 194-0004, Japan
ktotsu@ohtsu.jst.go.jp

Classical Theory on Electromagnetic Near Field

I. Banno

1 Introduction

This work is focused on the classical theory of the electromagnetic (EM) near field in the vicinity of matter. The EM near field is rather dependent on Maxwell's boundary conditions (MBCs). In a low symmetric system the MBCs cause difficulty in our understanding of the physics and in numerical calculations. In order to overcome this difficulty we develop two novel formulations, namely a boundary scattering formulation with scalar potential and a boundary scattering formulation with dual EM potential. Both the formulations are appropriate not only for carrying out numerical calculations but also to give an intuitive picture of the EM near field.

The motivation of our work is the next question: why is a resolution far beyond the diffraction limit, namely super-resolution, attained in near-field optical microscopy (NOM)? In this section, we review the experiments and the theory concerning NOM. Then the purposes and the overview of this chapter are given.

1.1 Studies of Pioneers

The first suggestion of a microscope with super-resolution appeared in a paper by Synge in 1928 [1]. The same idea was seen in a letter by O'Keefe in 1956 [2]. Synge's proposal is sketched in Fig. 1. A sample is placed on the true plane of glass and exposed to penetrating visible light through a small aperture. The size of the aperture and the distance between the sample and the aperture are much smaller than the wavelength of the visible light. A part of the penetrating light is scattered by the sample and reaches the photoelectric detector. By varying the position of the sample, one obtains the signal-intensity profile, that is, the electric current intensity as a function of the position of the sample. Synge pointed out the technical difficulty in his period and it has been overcome as time has progressed.

The first experiment of a microscope with super-resolution $\lambda/60$ was demonstrated in the microwave region in 1972 [3], then in the infrared region in 1985 with the resolution $\lambda/4$ [4]; λ stands for the wavelength of the EM field. The super-resolution in the optical region was attained by Pohl et al. in 1984 [5]; they implied that the resolution is $\lambda/20$. They formed a small

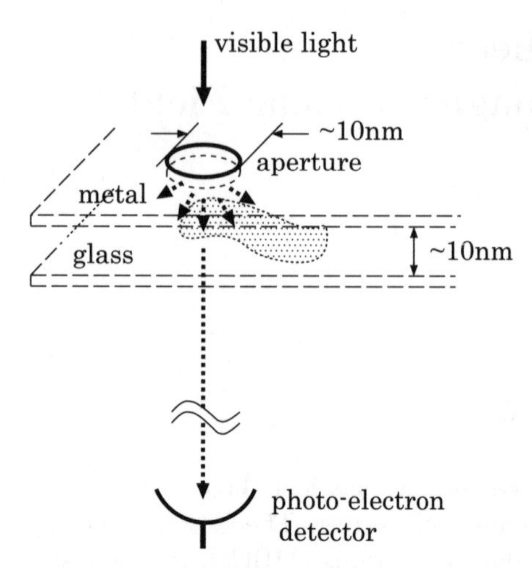

Fig. 1. A sketch of Synge's idea

aperture on the top of a metal-coated quartz tip; the radius of curvature of the sharpened tip was about 30 nm. Their result demonstrated microscopy with super-resolution in the visible light region. In 1987, Betzig et al. [6] attained super-resolution under "collection mode" in the visible light region. In the collection mode, the incident light exposes a wide region including the sample; the light scattered by the sample is picked up by an aperture on a metal-coated probe tip. They used visible light and an aperture with a diameter $\sim$ 100 nm. The first experiment with high reproducibility and with nanometer resolution was done in 1992, using an aperture with a diameter $\sim$ 10 nm [7,8].

For a long time, the theoretical approach for the EM near-field problem had been based on the diffraction theory for a high symmetric system [9,10]. After the collection-mode operation was made popular in the 1990s, the EM scattering theories were applied and various numerical calculations have been carried out in low symmetric systems. Some workers solved the Dyson equation followed by Green's function [11,12] and others calculated the time evolution of the EM field by the finite differential time domain (FDTD) method [13]. Both methods had been originally developed for the calculation of EM far field and have never produced an intuitive physical picture of the EM near field.

1.2 Purposes of This Chapter

The purposes of this chapter are:

1. To give a clear definition of far field and near field.
2. To calculate the EM near field on the basis of two novel formulations free from the MBCs, namely the boundary scattering formulation with scalar potential and that with dual EM potential.

3. To give a clear physical picture of EM near field on the basis of our formulations, eliminating the difficulty of the MBCs.

1.3 Overview of This Chapter

In Sect. 2, we will make clear the elementary concepts: retardation effect, diffraction limit, far field and near field. To understand them is a prerequisite to the subsequent sections.

The system of interest to us is characterized by the condition, $ka \lesssim kr \lesssim 1$, where "$a$" is the representative size of the matter, "r" is the distance between the matter and the observation point and "k" is the wavenumber of the incident EM field. Under this condition, the boundary effect – the effect of the MBCs – is relatively larger than (or comparable to) the retardation effect. Therefore it is crucial to determine how to treat the MBCs in a EM near-field problem. However, the boundary value problem in a low symmetric system is troublesome not only in a numerical calculation but also in the understanding physics.

To overcome this difficulty caused by the MBCs, we introduce two formulations based on the following principles:

1. The EM potential is the minimum degree of freedom of the EM field.
2. A boundary value problem can be replaced by a scattering problem with an adequate boundary source; this boundary source is responsible for the MBCs.

In Sect. 3, we will develop the boundary scattering formulation with scalar potential; this formulation is available under the "near-field condition"(NFC), i.e., $ka \lesssim kr \ll 1$. In this limiting case, the retardation effect is negligible and a quasistatic picture holds, that is, the static Coulomb law governs the electric field under the NFC. We can use the scalar potential as the minimum degree of freedom of the electric field. Furthermore, we can introduce an adequate boundary charge density to reproduce the MBCs. In this way, a boundary value problem under the NFC can be replaced by a scattering problem with an adequate boundary source, namely *a boundary scattering problem.* We can solve this problem for the scalar potential using a perturbative or an iterative method. The field distribution in the vicinity of a dielectric can be intuitively understood on the basis of the static Coulomb law. The boundary scattering formulation with the scalar potential is also applicable to a static electric problem and a static magnetic one.

In Sect. 4, we will derive the *dual EM potential* from the ordinary EM potential by means of dual transformation; dual transformation is the mutual exchange between the electric quantities and the magnetic ones. The dual EM potential in the radiation gauge is the minimum degree of freedom of the EM field under the condition that the magnetic response of the matter is negligible. The source of the dual EM potential is the magnetic current density and we can define an adequate boundary magnetic current density to reproduce the MBCs. In this way, we can replace a boundary value problem by a scattering problem with an adequate boundary source, namely *a boundary scattering problem.* The boundary scattering formulation with the dual EM potential is applicable to both the far-field problem and near-field one.

In Sect. 5, we will apply the boundary scattering formulation with the dual EM potential to the EM near field of a dielectric under $ka \lesssim kr \lesssim 1$; the boundary effect and the retardation effect coexist under this condition. We will numerically solve the boundary scattering problem for the dual EM potential and also give an intuitive understanding on the basis of the "dual Ampere law" with a correction due to the retardation effect.

In Sect. 6, we will give the summary of this chapter.

As the first stage of the investigation, all the numerical calculations in this chapter are restricted to the EM near field in the vicinity of a *dielectric*, although our formulations can be extended to treat various types of material, e.g., a metal, a magneto-optical material, a nonlinear material and so on.

There are three additional sections, Sects. 7–9, corresponding to appendices. Section 7 concerns formulas for the far-field intensity, the near-field intensity and the signal intensity in NOM. Sections 8 and 9 are mathematical details on boundary source and vector Green's function, respectively.

2 Definition of Near Field and Far Field

Although a certain simple property seems to exist in EM near field, a physical picture is smeared behind the complicated calculation procedures caused by the MBCs. There is no formalism on EM near field compatible with a clear physical picture. So, before a discussion on EM near field, let us reconsider wave mechanics in a general point of view and make clear the following concepts: the retardation effect, the diffraction limit, the far field and the near field [16][1].

2.1 A Naive Example of Super-Resolution

First, we introduce a simple example to explain why a super-resolution is attained in NOM. Suppose a small stone is thrown into a pond. One finds that circular wavelets extend on the surface of the water. Ensure that the shapes of the wavelets are circular independently of the shape of the stone. This means that we cannot know the shape of the stone, if the observation points are far from the source point. Strictly speaking, there is a "diffraction limit" in the far-field observation, if the size of stone is much smaller than the wavelength of the surface wave of the water. However, information concerning

[1] In the above review (in Japanese) the numerical results in Fig. 4 and the related discussion are incorrect; those concern the retardation effect in the vicinity of a dielectric. This error is modified in this chapter, see Sect. 5.4.

the shape of the stone is not lost if an observation point is close enough to the source point. This type of observation is just that in NOM. In Sect. 2.2, this idea will be developed using a pedagogical model.

2.2 Retardation Effect as Wavenumber Dependence

Suppose that there are two point sources instead of a complicated-shaped source like a stone, see Fig. 2. These sources yield a scalar field and are located at $\boldsymbol{r}' = +\boldsymbol{a}/2$ and $\boldsymbol{r}' = -\boldsymbol{a}/2$ in three-dimensional space. Furthermore, we assume that the two sources oscillate with the same phase and the same magnitude, i.e., $\delta^3(\boldsymbol{r}' \pm \boldsymbol{a}/2)\exp(\mathrm{i}\omega t')$, where ω is the angular frequency. Our simplified problem is to know $a = |\boldsymbol{a}|$ by means of observation at some points $\boldsymbol{r}'$s. It is assumed that the directional vector $\hat{\boldsymbol{a}}$ is known. Let us consider the front of the wave that starts from each source point $\boldsymbol{r}' = \mp\boldsymbol{a}/2$ at the time $t' = 0$. The front of each wave reaches the observation point "$\boldsymbol{r}$" at a certain time $\Delta t_{\mp\boldsymbol{a}/2}$. This time – "retardation" – is needed for the wave to propagate from $\boldsymbol{r}' = \mp\boldsymbol{a}/2$ to $\boldsymbol{r}$ with the phase velocity ω/k. Therefore, the retardation is estimated as

$$\Delta t_{\mp\boldsymbol{a}/2} = k|\boldsymbol{r} \pm \boldsymbol{a}/2|/\omega\ . \tag{1}$$

After all, the amplitude of each partial wave at the observation point $(\boldsymbol{r}, t)$ is the following,

$$\frac{\exp(-\mathrm{i}\omega(t - \Delta t_{\mp\boldsymbol{a}/2}))}{|\boldsymbol{r} \pm \boldsymbol{a}/2|} = \frac{\exp(-\mathrm{i}\omega t + \mathrm{i}k|\boldsymbol{r} \pm \boldsymbol{a}/2|)}{|\boldsymbol{r} \pm \boldsymbol{a}/2|}\ . \tag{2}$$

The magnitude of each partial wave is in inverse proportion to the distance between the observation point and the source point because of the conservation of flux. The phase is just that at the source point in the time $t - \Delta t_{\mp\boldsymbol{a}/2}$

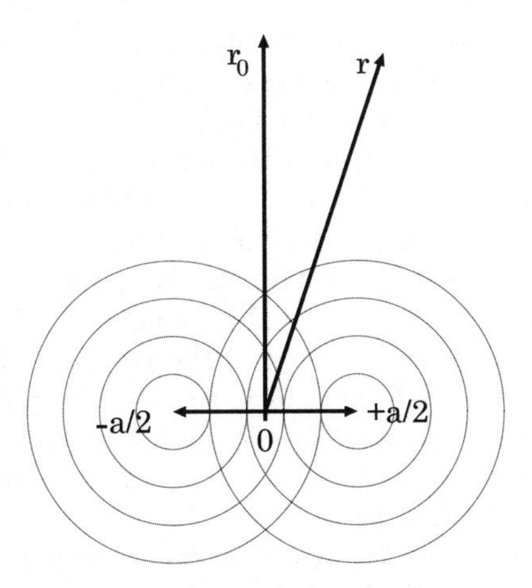

Fig. 2. A system with two point sources. Only one parameter $a = |\boldsymbol{a}|$ characterizes "the shape of the source" if $\hat{a}$ is given

because of the retardation. Equation (2) is an expression for Huygens' principle or Green's function for the scalar Helmholtz equation.

The expression (1) for the retardation tells us that *the retardation effect is the wavenumber dependence*, namely ka- and/or kr-dependence. In the following, the retardation effect will be used in this meaning.

2.3 Examination on Three Cases

In order to explain the meaning of the diffraction limit and to give a clear definition of far field and near field, let us discuss the next three cases:

- case 1; $kr \gg ka \gg 1$, the observation of the far field yielded by a large-sized source,
- case 2; $kr \gg 1 \gg ka$, the observation of the far field yielded by a small-sized source,
- case 3; $ka \lesssim kr \ll 1$, the observation of the near field yielded by a small-sized source.

In our simplified model introduced in Sect. 2.2, the observed amplitude at $(\boldsymbol{r}, t)$ is the superposition of the two partial waves,

$$A(\boldsymbol{r}, t) = \frac{\exp(-\mathrm{i}\omega t + \mathrm{i}k|\boldsymbol{r} + \boldsymbol{a}/2|)}{|\boldsymbol{r} + \boldsymbol{a}/2|} + \frac{\exp(-\mathrm{i}\omega t + \mathrm{i}k|\boldsymbol{r} - \boldsymbol{a}/2|)}{|\boldsymbol{r} - \boldsymbol{a}/2|} . \tag{3}$$

In cases 1 and 2, (3) is reduced to the next expression, applying the condition $r \gg a$,

$$A(\boldsymbol{r}, t) = \frac{e^{-\mathrm{i}\omega t + \mathrm{i}kr}}{r} \left(2\cos\left(\frac{1}{2}ka(\hat{\boldsymbol{r}} \cdot \hat{\boldsymbol{a}})\right) + \mathcal{O}\left(\frac{a}{r}\right)\right) . \tag{4}$$

In (4), the "a" in question is coupled with "k". Therefore one must use the retardation effect (in the phase difference between the two partial waves) to determine "a" by means of the far-field observation.

In case 1, "a" can be obtained in the following way. We restrict the observation points on a sphere $r = \text{const.}$ ($\gg a$) and select one point $\boldsymbol{r}_0$ on the sphere that satisfies $\hat{\boldsymbol{r}}_0 \cdot \hat{\boldsymbol{a}} = 0$. At this observation point, the phase difference of the two partial waves is 0. Then, find another point $\boldsymbol{r}$ on the sphere where the magnitude of the field A takes local minimum. If $\boldsymbol{r}$ is one of the nearest points of $\boldsymbol{r}_0$, the phase difference at $\boldsymbol{r}$ of the two partial waves is $\pi/2$, i.e., $ka|\hat{\boldsymbol{r}} \cdot \hat{\boldsymbol{a}}|/2 = \pi/2$. As a result, we determine "a"; $a = \pi/(k|\hat{\boldsymbol{r}} \cdot \hat{\boldsymbol{a}}|)$.

In case 2, however, "a" cannot be obtained through the far-field observation. The phase difference among all the observation points on the sphere is zero because of the condition $ka \ll 1$. In other words, the two point sources are so close that the observer far from the sources recognizes the two sources as a single source with the double magnitude.

In case 3, (3) is reduced to the next expression, applying the condition $ka \lesssim kr \ll 1$,

$$A(\boldsymbol{r}, t) = \exp(-\mathrm{i}\omega t)\left(\frac{1}{|\boldsymbol{r}+\boldsymbol{a}/2|}+\frac{1}{|\boldsymbol{r}-\boldsymbol{a}/2|}\right)(1+\mathcal{O}(ka, kr)) \;. \quad (5)$$

The leading order of (5) is independent of the wavenumber "k". This independence is because the size of the whole system – including all the sources and all the observation points – is much smaller than the wavelength, that is, the system cannot feel the wavenumber. The "a" in question can be determined by means of the near-field observation. Find an observation point $\boldsymbol{r}_0$ where $\hat{\boldsymbol{r}}_0 \cdot \hat{\boldsymbol{a}} = 0$ is satisfied, then (5) results in $a = 2\sqrt{4 - r_0^2|A_0|^2}/|A_0|$. Make sure that this expression for "a" is independent of "k", i.e., independent of the retardation effect. As a result, we can know "a" through the near-field observation without using the retardation effect.

In short, information concerning the shape of the source is in the k-dependent phase of the far field or in the k-independent magnitude of the near field.

2.4 Diffraction Limit in Terms of Retardation Effect

In the above pedagogical model, the scalar field yielded by the two point sources has been discussed. Even in the case of a continuous source, the essential physics is the same, if "a" is considered as the representative size of the source.

Now we can make clear the meaning of the diffraction limit. In the case of a far-field observation, information concerning the shape of the source is in the k-dependent phase of the far field, see (4). To recognize the anisotropy of the shape, the phase difference among some observation points on a certain sphere must be larger than $\pi/2$; this condition imposed on (4) leads to the next inequality,

$$ka|\hat{\boldsymbol{r}} \cdot \hat{\boldsymbol{a}}| \sim ka \gtrsim \pi \;, \quad (6)$$

where "a" is the representative size of the source. The inequality (6) is a rough expression for the diffraction limit and implies that the size of the source should be larger than the order of the wavelength to detect the anisotropy of the source. Note that the concept "diffraction limit" is effective only in the far-field observation, i.e., the observation under the condition $kr \gg 1,\ r \gg a$.

In the far-field observation like cases 1 and 2, we have to use the k-dependent phase to know the shape of the source and the resolution is bounded by the diffraction limit, see (4) and (6). However, in the near-field observation like case 3, we know the shape of the source without diffraction limit because information of the shape is in the k-independent magnitude of the near field, see (5).

Table 1. Definition and specification of near field and far field

Definition	Diffraction limit	Retardation effect	Examples
Far field	Exists	Large	$ka \gg 1$ (case 1) ordinary optical microscopy
$kr \gg 1, r \gg a$			$ka \ll 1$ (case 2) Rayleigh's phenomena
Near field $1 \gtrsim kr \gtrsim ka$	Does not exist	Small	$ka \ll 1$ (case 3) NOM

2.5 Definition of Near Field and Far Field

The observation in NOM corresponds to case 3, if the position of the probe tip is considered as the observation point. In fact, the signal in NOM is independent of the wavenumber and free from the diffraction limit. On the contrary, the observation in the usual optical microscopy corresponds to case 1, therefore, the resolution in it is bounded by the diffraction limit. Case 2 is the condition for Rayleigh's scattering phenomena. By means of a far-field observation, one can determine only the number (or density) of the sources as a whole but cannot obtain information about the shape or the distribution of the sources.

We define "far field" as the field observed under the condition $kr \gg 1$, $r \gg a$, i.e., cases 1 and 2, and "near field" as the field observed under the condition $ka \lesssim kr \lesssim 1$, i.e., case 3.

In particular, the limiting condition of the near field,

$$ka \lesssim kr \ll 1 \, , \tag{7}$$

is simply referred as the "near-field condition (NFC)" in the following. Under the NFC, the k-independent picture, namely a quasistatic picture, holds owing to the negligible retardation effect, see Sect. 3.

In this section, we treat only a scalar field. A quasistatic picture is also effective for the EM near field, although EM field is a vector field. It is a characteristic of the EM near field that the k-independent boundary effect is dominant; this will be apparent in Sect. 5.1.

3 Boundary Scattering Formulation with Scalar Potential

In this section, a formulation to treat the electric field under the NFC (7) is given. Under the NFC, a quasistatic picture holds and the scalar potential is the minimum degree of freedom of the EM field. Furthermore, we replace the boundary value problem by *a boundary scattering problem*, i.e., a scattering problem with an adequate boundary source. The boundary scattering formulation with the scalar potential gives an intuitive picture of the near field in the vicinity of a dielectric and a simple procedure of numerical calculation [16].

3.1 Quasistatic Picture under Near-Field Condition

Suppose that a three-dimensionally small piece of matter with linear response is exposed to an incident EM field; we observe the EM field in the vicinity of the matter. The following notations are introduced: "a" stands for the representative size of the matter, $\boldsymbol{k}$ and $\boldsymbol{E}^{(0)}$ are the wavenumber vector and the polarization vector of the incident light respectively, and $\boldsymbol{r}$ is the position vector of the observation point relative to the center of the matter.

We assume the NFC (7), that is, all the retardation effects ($\boldsymbol{k}$-dependence) in Maxwell's equations are negligible and the quasistatic ($\boldsymbol{k}$-independent) picture holds. Figure 3 is a snapshot of such a system at an arbitrary time. The magnetic field is the incident field itself because a negligible magnetic response of the matter is assumed. Therefore, we concentrate ourselves on the electric field. Similarly to the electrostatic field, the electric near field under the NFC is derived from the scalar potential,

$$\boldsymbol{E}(\boldsymbol{r})\exp(-\mathrm{i}\omega t) = -\nabla\phi(\boldsymbol{r})\exp(-\mathrm{i}\omega t), \tag{8}$$

where ω is the angular frequency of the field. Note that the value of ω is that of the incident field because of the linear response of the matter. From (8) the electric near field under the NFC is a longitudinal field, i.e., a nonradiative field. This fact strikingly contrasts with the fact that a radiative field in the

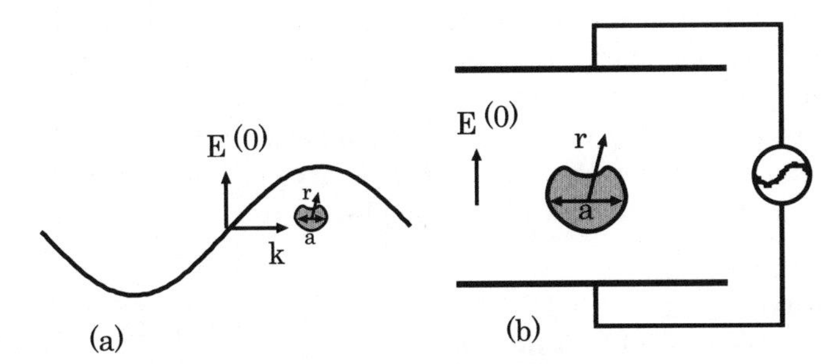

Fig. 3a,b. A quasistatic picture under the NFC. (**a**) A snapshot of a system under the NFC at an arbitrary time. The matter, which is much smaller than the wavelength, is exposed to the incident field with the wavenumber vector $\boldsymbol{k}$ and the polarization vector $\boldsymbol{E}^{(0)}$. (**b**) An equivalent quasistatic system under an alternating voltage

far-field regime is transversal. Therefore, we discuss the electric field under the NFC as the first step and extract the properties of the EM near field.

In the following, we will omit the common time-dependent factor, $\exp(-\mathrm{i}\omega t)$ for simplicity. The real electric field at the time "t" can be calculated by $\mathcal{R}e\{\boldsymbol{E}(\boldsymbol{r})\exp(-\mathrm{i}\omega t)\}$.

3.2 Poisson's Equation with Boundary Charge Density

Under the NFC, Maxwell's equations are reduced to the static Coulomb law

$$\nabla \cdot \boldsymbol{E}(\boldsymbol{r}) = -\frac{1}{\epsilon_0}\nabla \cdot \boldsymbol{P}(\boldsymbol{r}) \,, \tag{9}$$

$$\boldsymbol{P}(\boldsymbol{r}) = (\epsilon(\boldsymbol{r}) - \epsilon_0)\boldsymbol{E}(\boldsymbol{r}) \,, \tag{10}$$

where $\boldsymbol{P}(\boldsymbol{r})$ is the polarization defined by the local and linear dielectric function $\epsilon(\boldsymbol{r})$; $\epsilon(\boldsymbol{r})$ is assumed to be a smooth function of $\boldsymbol{r}$ for a time.

In terms of the scalar potential, which is defined as $\boldsymbol{E}(\boldsymbol{r}) = -\nabla\phi(\boldsymbol{r})$, (9) and (10) become Poisson's equation,

$$-\triangle\phi(\boldsymbol{r}) = \nabla \cdot \left(\frac{\epsilon(\boldsymbol{r}) - \epsilon_0}{\epsilon_0}\nabla\phi(\boldsymbol{r})\right) \,. \tag{11}$$

Now, using $\nabla \cdot \{\epsilon(\boldsymbol{r})\nabla\phi(\boldsymbol{r})\} = \nabla\epsilon(\boldsymbol{r}) \cdot \nabla\phi(\boldsymbol{r}) + \epsilon(\boldsymbol{r})\triangle\phi(\boldsymbol{r})$, (11) results in another expression for Poisson's equation,

$$-\triangle\phi(\boldsymbol{r}) = \frac{\nabla\epsilon(\boldsymbol{r})}{\epsilon(\boldsymbol{r})} \cdot \nabla\phi(\boldsymbol{r}) \,. \tag{12}$$

The r.h.s. of (12) is the induced charge density divided by ϵ_0; this charge density is localized within the interface region, where $\epsilon(\boldsymbol{r})$ varies steeply and $\nabla\epsilon(\boldsymbol{r})$ takes a large value. In the following, we simply refer to the quantity in the r.h.s. of (12) as the "boundary charge density", ignoring the constant factor $1/\epsilon_0$. The "boundary charge density" means the charge per unit volume (not per unit area) in this chapter.

Equations (11) and (12) are equivalent but we prefer (12), which is the starting point of our novel formulation and gives a simple procedure of numerical calculation together with a clear physical picture.

3.3 Intuitive Picture of EM Near Field under Near-Field Condition

Suppose that the piece of matter is a dielectric with a steep interface. On the basis of (12), one can obtain an intuitive picture of the electric field near the dielectric under the NFC.

Let us consider (12) in the limit of the steep interface. $\nabla\epsilon(\boldsymbol{r})$ leads to $-(\epsilon_1 - \epsilon_0)\delta(\boldsymbol{r} \in \text{boundary})\boldsymbol{n}_s$, where $\boldsymbol{n}_s$ stands for the outward directional vector normal to the boundary between the matter and the vacuum, and $\delta(\boldsymbol{r} \in \text{boundary})$ is the one-dimensional delta function in the direction of $\boldsymbol{n}_s$. Furthermore, $\nabla\phi(\boldsymbol{r}) = -\boldsymbol{E}(\boldsymbol{r}) \simeq -\boldsymbol{E}^{(0)}$ (const.) under $|\Delta\boldsymbol{E}| \ll |\boldsymbol{E}^{(0)}|$, where $\Delta\boldsymbol{E}$ is the scattered field defined by $\Delta\boldsymbol{E} \equiv \boldsymbol{E} - \boldsymbol{E}^{(0)}$. Note that under the NFC, the incident field is regarded as the constant vector $\boldsymbol{E}^{(0)}$ over the whole system, see Fig. 3. Then, using $\nabla \cdot \boldsymbol{E}^{(0)} = 0$, (12) results in

$$\nabla \cdot \Delta\boldsymbol{E}(\boldsymbol{r}) = \delta(\boldsymbol{r} \in \text{boundary})\frac{\epsilon_1 - \epsilon_0}{\epsilon(\boldsymbol{r})}\boldsymbol{n}_s \cdot \boldsymbol{E}^{(0)} . \tag{13}$$

Here the value of $\epsilon(\boldsymbol{r})$ in the denominator is not well defined but a value between ϵ_0 and ϵ_1 is physically acceptable in a naive sense. If the shape of the dielectric is isotropic enough, like a sphere, we recommend to set $\epsilon(\boldsymbol{r}) = 1/3\epsilon_1 + 2/3\epsilon_0$, see Sect. 3.8. Equation (13) is fully justified in Sect. 3.6.

Once the boundary source is estimated, one can intuitively imagine the electric flux or the scattered field $\Delta\boldsymbol{E}(\boldsymbol{r})$ making use of the Coulomb law (13). Figure 4 describes such a relation between the boundary source and the electric flux. Furthermore, we can simply interpret the electric field intensity. The time-averaged intensity at the position $\boldsymbol{r}$ is defined as (14) in terms of the complex electric field.

$$\begin{aligned} \Delta I(\boldsymbol{r}) &= \frac{|\boldsymbol{E}(\boldsymbol{r})|^2 - |\boldsymbol{E}^{(0)}(\boldsymbol{r})|^2}{|\boldsymbol{E}^{(0)}(\boldsymbol{r})|^2} \\ &= \frac{\boldsymbol{E}^{(0)*}(\boldsymbol{r}) \cdot \Delta\boldsymbol{E}(\boldsymbol{r}) + \text{c.c.} + |\Delta\boldsymbol{E}(\boldsymbol{r})|^2}{|\boldsymbol{E}^{(0)}(\boldsymbol{r})|^2} , \end{aligned} \tag{14}$$

where $|\boldsymbol{E}^{(0)}|^2$ in the denominator is introduced to make ΔI dimensionless and that in the numerator is done to eliminate the background intensity. $\Delta I(\boldsymbol{r}) > 0[< 0]$ means that the intensity at $\boldsymbol{r}$ is larger (smaller) than that of the background. If $\epsilon_1/\epsilon_0 \sim 1$, $\Delta\boldsymbol{E}(\boldsymbol{r})$ is so small that $\Delta I(\boldsymbol{r})$ is determined by the inner product of $\boldsymbol{E}^{(0)}$ and $\Delta\boldsymbol{E}(\boldsymbol{r})$. For example, in the system of Fig. 4, the negative intensity $\Delta I(\boldsymbol{r}) < 0$ appears at the observation points over the top of the matter. This is because $\boldsymbol{E}^0$ and $\Delta\boldsymbol{E}(\boldsymbol{r})$ are antiparallel at such observation points.

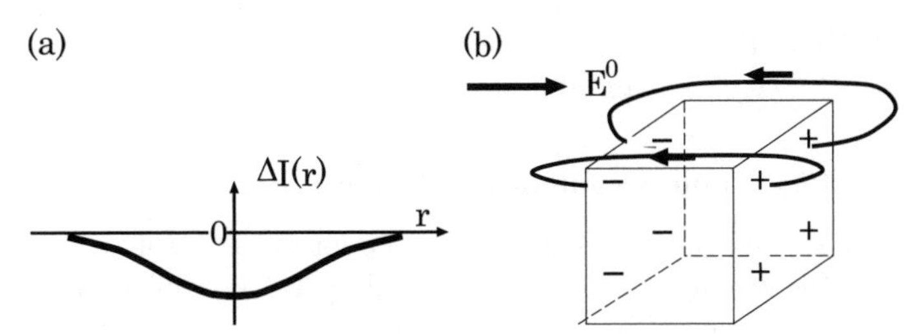

Fig. 4a,b. An intuitive picture of the EM near field of a dielectric under the NFC. (**a**) The profile of electric field intensity along a scanning line parallel to $\boldsymbol{E}^0$ over the matter. (**b**) The electric flux yielded by the induced boundary charge

Even if the shape of a dielectric is complicated, the above procedure to understand $\Delta\boldsymbol{E}(\boldsymbol{r})$ and $\Delta I(\boldsymbol{r})$ is available, see Sect. 3.10.

Now we have used (14) for the formula for the near-field intensity but this intensity itself is not considered as the signal intensity in NOM. Furthermore, a formula for the far-field intensity is different from that of the near-field intensity. These three formulas are discussed in Sect. 7.

3.4 Notations Concerning Steep Interface

In Sect. 3.3, we have applied (12) to a system with a steep interface in a rather rough manner. Strictly speaking, there is a difficulty to treat (12) in the limit of the steep interface, that is, the boundary charge density is a product of distributions and not well defined in a general sense of mathematics. Actually, the quantities $\epsilon(\boldsymbol{r})$, $\nabla\phi(\boldsymbol{r})$, and $\nabla\epsilon(\boldsymbol{r})$ in (12) become distributions, i.e., the step function and/or the delta function, in the limit of the steep interface.

To treat this singularity in a proper manner, let us introduce some notations. A steep interface is characterized by a stepwise dielectric function,

$$\epsilon(\boldsymbol{r}) \equiv \epsilon_0 + \theta(\boldsymbol{r} \in \mathcal{V}_1)(\epsilon_1 - \epsilon_0) \,, \tag{15}$$

$$\theta(\boldsymbol{r} \in \mathcal{V}_1) \equiv \begin{cases} 0 & \text{for } \boldsymbol{r} \in \mathcal{V}_0 \\ \text{not defined} & \text{for } \boldsymbol{r} \in \mathcal{V}_{01} \\ 1 & \text{for } \boldsymbol{r} \in \mathcal{V}_1 \end{cases} \,, \tag{16}$$

where $\mathcal{V}_0$, $\mathcal{V}_1$, and $\mathcal{V}_{01}$ stand for the vacuum, the matter, and the interface region, respectively, and ϵ_1 stands for the complex dielectric constant of the matter. Make sure that $\mathcal{V}_{01}$ is volume, i.e., the three-dimensional, space with infinitesimal width $\eta = +0$. A definition of $\epsilon(\boldsymbol{r})$ in the interface region is not given because it is not needed in the following discussion.

We give the next notations (Fig. 5): $\mathcal{V}_{01}/\eta$ is the whole boundary of the matter (two-dimensional space), $\boldsymbol{s} \in \mathcal{V}_{01}/\eta$ is a position vector on the boundary located in the center of $\mathcal{V}_{01}$, $\boldsymbol{n}_s$ is the outward normal vector at $\boldsymbol{s}$, $\sigma \subset \mathcal{V}_{01}/\eta$ is the small boundary element containing $\boldsymbol{s}$, and $\boldsymbol{s}_0$ and $\boldsymbol{s}_1$ are the position vectors just outside the interface region defined as $\boldsymbol{s}_0 \equiv \boldsymbol{s} + \frac{\eta}{2}\boldsymbol{n}_s = \boldsymbol{s} + 0\boldsymbol{n}_s$ and $\boldsymbol{s}_1 \equiv \boldsymbol{s} - \frac{\eta}{2}\boldsymbol{n}_s = \boldsymbol{s} - 0\boldsymbol{n}_s$. Further, $\boldsymbol{m}_1$ and $\boldsymbol{m}_2$ are two independent unit vectors in the boundary at $\boldsymbol{s}$, l_i $(i = 1, 2)$ is a small length along $\boldsymbol{m}_i$ $(i = 1, 2)$ (so that $\sigma = l_1 \otimes l_2$), $\eta \otimes l_i$ $(i = 1, 2)$ is an infinitesimal area and $\sigma \otimes \eta = l_1 \otimes l_2 \otimes \eta$ is an infinitesimal volume.

3.5 Boundary Value Problem for Scalar Potential

To overcome the difficulty caused by the steep interface, a well-known means is to replace the original problem by a boundary value problem. In our context, (12) can be replaced by the next equations,

$$-\Delta\phi(\boldsymbol{r}) = 0 \qquad \text{for } \boldsymbol{r} \in \mathcal{V}_0 \cup \mathcal{V}_1 \,, \tag{17a}$$

$$\epsilon_0 \partial_n \phi(\boldsymbol{s}_0) = \epsilon_1 \partial_n \phi(\boldsymbol{s}_1) \qquad \text{for } \boldsymbol{s} \in \mathcal{V}_{01}/\eta \,, \tag{17b}$$

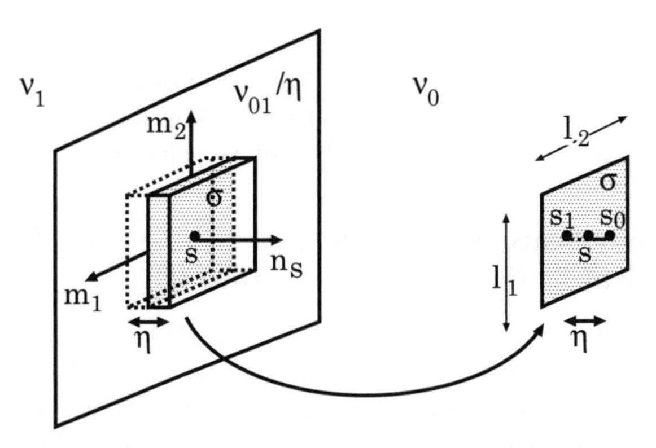

Fig. 5. Notations concerning a steep interface

where $\partial_n \phi \equiv \boldsymbol{n}_s \cdot \nabla \phi$. The boundary condition (17b) describes the discontinuity of the boundary-normal component of the electric field. Equation (17b) is derived from (12) as follows. Keeping η finite and assuming that $\epsilon(\boldsymbol{r})$ in the interface region is smooth enough, (12) is equivalent to

$$\nabla \cdot \left(\frac{\epsilon(\boldsymbol{r})}{\epsilon_0} \nabla \phi(\boldsymbol{r}) \right) = 0 . \tag{18}$$

R.P. Feynmann pointed out that this type of equation appears in various fields of physics [17]. Integrating (18) over the small volume $\sigma \otimes \eta$ in Fig. 5, and applying Gauss' theorem and taking the limit $\eta \to +0$, one obtains (17b). Make sure that an explicit formula for $\epsilon(\boldsymbol{r})$ in the interface region is not needed in the above derivation of (17b). That is, the MBC (17b) is independent of a dielectric function in the interface region.

Outside the interface region, i.e., in $\mathcal{V}_0 \cup \mathcal{V}_1$, the solution of (12) under (15) is obtained by solving (17a)–(17b) of the boundary value problem. In the boundary value problem, the boundary condition (17b) contains sufficient information to construct the solution outside the interface region, therefore one does not require the source or the field in the interface region. See Sect. 8 for a detailed discussion starting from a given dielectric function in the interface region.

Note that we may know one more MBC concerning the electric field; it describes the continuity of the boundary-parallel component of the electric field,

$$\boldsymbol{n}_s \times \nabla \phi(\boldsymbol{s}_0) = \boldsymbol{n}_s \times \nabla \phi(\boldsymbol{s}_1) . \tag{19}$$

Equation (19) is trivial because it is derived from the identity $\nabla \times \nabla \phi(\boldsymbol{r}) = 0$. Integrating this identity over the small area $l_i \otimes \eta$ for $i = 1, 2$, applying Stokes' theorem and taking the limit $\eta \to +0$, one obtains (19). Therefore, we do not need a boundary condition (19) in the calculation in terms of the scalar potential.

It is troublesome to solve a boundary value problem in a low symmetric system. Even if one can obtain a solution fortunately, it is still difficult to give a clear physical picture. In the next subsection, we will propose another way free from these difficulties in the boundary value problem.

3.6 Boundary Scattering Problem Equivalent to Boundary Value Problem

To overcome the difficulty to solve (12) under (15), i.e., in a system with the steep interface, there is another way, the *boundary scattering formulation* with the scalar potential. The difficulty in the original problem is that the boundary charge density in (12) becomes a product of distributions, which is not well defined in a general sense of mathematics. However, detailed analysis in Sect. 8 reveals that the boundary charge density is a well-defined quantity and can be expressed in various ways.

Using one of the possible expressions for the boundary charge density, (12) becomes (20a)–(20c), which are the elementary equations for the boundary scattering formulation with scalar potential,

$$-\Delta\phi(\boldsymbol{r}) = -\int_{\mathcal{V}_{01}/\eta} \mathrm{d}^2 s \delta^3(\boldsymbol{r}-\boldsymbol{s}) \frac{\epsilon_1 - \epsilon_0}{\epsilon(\boldsymbol{s})} \partial_n \phi(\boldsymbol{s}) \text{ for } \boldsymbol{r} \in \mathcal{V}_0 \cup \mathcal{V}_{01} \cup \mathcal{V}_1 \ , \quad (20\text{a})$$

$$\epsilon(\boldsymbol{s}) \equiv \alpha(\boldsymbol{s})\epsilon_1 + (1-\alpha(\boldsymbol{s}))\epsilon_0 \ , \quad (20\text{b})$$

$$\partial_n\phi(\boldsymbol{s}) \equiv (1-\alpha(\boldsymbol{s}))\partial_n\phi(\boldsymbol{s}_1) + \alpha(\boldsymbol{s})\partial_n\phi(\boldsymbol{s}_0) \ . \quad (20\text{c})$$

Equations (20b)–(20c) are merely the definitions of $\epsilon(\boldsymbol{s})$ and $\partial_n\phi(\boldsymbol{s})$, respectively; $\alpha(\boldsymbol{s})$ is an arbitrary smooth and complex-valued function on $\mathcal{V}_{01}/\eta$.

Here we only show that (20a)–(20c) of the boundary scattering problem leads to (17a)–(17b) of the boundary value problem. The derivation of (17a) is trivial and that of (17b) is as follows. Integrating (20a) over the infinitesimal volume element $\sigma \otimes \eta$ ($\eta = +0$) in Fig. 5, applying Gauss' theorem to the l.h.s. and carrying out the volume integral of the delta function in the r.h.s., then, it is confirmed that the MBC (17b) is reproduced from (20a)–(20c). The arbitrary function $\alpha(\boldsymbol{s})$ disappears automatically and does not affect the field in $\mathcal{V}_0 \cup \mathcal{V}_1$.

In principle, the solution of (17a)–(17b) of the boundary value problem and that of (20a)–(20c) of the boundary scattering problem are equivalent in the domain $\mathcal{V}_0 \cup \mathcal{V}_1$. Because both the solutions in the regions $\mathcal{V}_0 \cup \mathcal{V}_1$ satisfy the same boundary condition. However, (20a)–(20c) possess considerable merits compared with (17a)–(17b). The first merit is that the boundary effect can be treated by a perturbative or an iterative method, see Sect. 3.7. The second is that the arbitrariness of the expression for the boundary source can be used to control the convergence in a numerical calculation.

The arbitrariness in (20a)–(20c) comes from the degrees of freedom of the charge-density profile inside the interface region $\mathcal{V}_{01}$; the integrated charge density over the width η ($= +0$), not the detailed profile of the charge density, determines the field in $\mathcal{V}_0 \cup \mathcal{V}_1$. This integrated quantity is analogous to the well-known multipole moment; only the multipole moments of a source determine the EM field outside of the finite source region [18]. Therefore, it is reasonable that a certain arbitrariness of the boundary source appears. See Sect. 8 for the mathematical details of the boundary scattering formulation; the boundary charge density is a well-defined quantity and is expressed with arbitrariness.

As a result, the MBCs can be built into the definition of the boundary charge density and the boundary value problem based on (17a)–(17b) or the original problem based on (12) with (15) is replaced to the boundary scattering problem based on (20a)–(20c). The boundary scattering formulation enables a perturbative treatment of the boundary effect.

3.7 Integral Equation for Source and Perturbative Treatment of MBC

Equations (20a)–(20c) are converted to the next integral equation,

$$\phi(\boldsymbol{r}) = \phi^{(0)}(\boldsymbol{r}) + \int_{\mathcal{V}_{01}/\eta} \mathrm{d}^2 s G(\boldsymbol{r}, \boldsymbol{s}) \frac{\epsilon_1 - \epsilon_0}{\epsilon(\boldsymbol{s})} \partial_n \phi(\boldsymbol{s}) \,, \tag{21}$$

$$G(\boldsymbol{r}, \boldsymbol{r}_1) = \frac{-1}{4\pi} \frac{1}{|\boldsymbol{r} - \boldsymbol{r}_1|} \,, \tag{22}$$

where $\phi^{(0)}(\boldsymbol{r}) = -\boldsymbol{E}^{(0)} \cdot \boldsymbol{r}$ is the incident scalar potential; $\phi^{(0)}$ leads to the incident electric field $\boldsymbol{E}^{(0)} = -\nabla\phi^{(0)}(\boldsymbol{r})$. $G(\boldsymbol{r}, \boldsymbol{r}_1)$ is Green's function for the Laplace equation; it satisfies, $-\triangle G(\boldsymbol{r}, \boldsymbol{r}_1) = -\delta^3(\boldsymbol{r} - \boldsymbol{r}_1)$.

Equation (21) leads to an integral equation for the boundary source $\zeta(\boldsymbol{s}) \equiv -(\epsilon_1 - \epsilon_0)/\epsilon(\boldsymbol{s})\partial_n \phi(\boldsymbol{s})$, which is charge per unit area.

$$\begin{aligned} \zeta(\boldsymbol{s}) = \; & \zeta^{(0)}(\boldsymbol{s}) + \frac{\epsilon_1 - \epsilon_0}{\epsilon(\boldsymbol{s})} \int_{\mathcal{V}_{01}/\eta} \mathrm{d}^2 s' \left\{ (1 - \alpha(\boldsymbol{s}))\, \boldsymbol{n}_s \cdot \nabla_{s_1} G(\boldsymbol{s}_1, \boldsymbol{s}') \right. \\ & \left. + \alpha(\boldsymbol{s}) \boldsymbol{n}_s \cdot \nabla_{s_0} G(\boldsymbol{s}_0, \boldsymbol{s}') \right\} \zeta(\boldsymbol{s}') \quad \text{for } \boldsymbol{s} \in \mathcal{V}_{01}/\eta \,, \end{aligned} \tag{23a}$$

$$\zeta^{(0)}(\boldsymbol{s}) = -\frac{\epsilon_1 - \epsilon_0}{\epsilon(\boldsymbol{s})} \partial_n \phi^{(0)}(\boldsymbol{s}) = \frac{\epsilon_1 - \epsilon_0}{\epsilon(\boldsymbol{s})} \boldsymbol{n}_s \cdot \boldsymbol{E}^{(0)} \,. \tag{23b}$$

In a numerical calculation based on the boundary scattering formulation with the scalar potential, the essential step is to solve (23a)–(23b). Once we obtain the source ζ, we can easily calculate the electric field using (21) together with $\boldsymbol{E} = -\nabla\phi$.

The usual perturbative method can be applied to (23a)–(23b) and the solution satisfies the MBCs to a certain degree, according to the order of approximation.

After all, our formulation to calculate the EM near field under the NFC is free from MBCs and one can treat the boundary effect using a perturbative or an iterative method.

3.8 Application to a Spherical System: Analytical Treatment

As an instructive application of the boundary scattering formulation, let us solve (23a)–(23b) analytically for a spherical symmetric system and examine how the arbitrariness in our formulation works in a perturbative or an iterative method. Let us consider the electric near field under the NFC in a system with a spherical piece of matter; the radius is "a'" and the sphere is exposed to the incident EM field. We do not need to restrict the value ϵ_1/ϵ_0 in the analytical calculation of this subsection. The notations are given in Sect. 3.4. For simplicity, we assume that $\alpha(\boldsymbol{s}) = \alpha$ (const.) for any $\boldsymbol{s}$.

The 0th-order boundary source $\zeta^{(0)}(\boldsymbol{s})$ is (24). Substituting this into $\zeta(\boldsymbol{s}')$ in the r.h.s. of (23a) leads to the 1st-order boundary source $\zeta^{(1)}(\boldsymbol{s})$ and so on. The n-th order source based on (23a)–(23b) is expressed by the geometrical series (25) and finally the infinite series is converted to the closed form (26),

$$\zeta^{(0)}(\boldsymbol{s}) = \frac{\epsilon_1 - \epsilon_0}{\alpha\epsilon_1 + (1-\alpha)\epsilon_0}\boldsymbol{n}_s \cdot \boldsymbol{E}^{(0)} , \tag{24}$$

$$\zeta^{(n)}(\boldsymbol{s}) = \sum_{k=0}^{n} \zeta^{(0)}(\boldsymbol{s}) \left(\frac{-1}{3} \frac{\epsilon_1 - \epsilon_0}{\alpha\epsilon_1 + (1-\alpha)\epsilon_0}(1-3\alpha) \right)^k , \tag{25}$$

$$\zeta^{(\infty)}(\boldsymbol{s}) = \frac{\epsilon_1 - \epsilon_0}{\frac{1}{3}\epsilon_1 + \frac{2}{3}\epsilon_0}\boldsymbol{n}_s \cdot \boldsymbol{E}^{(0)} = \zeta^{(0)}(\boldsymbol{s})|_{\alpha=1/3} . \tag{26}$$

We note some points:

- The α-dependence in ζ disappears in the rigorous solution (26), although the dependence remains under every finite-order approximation, see (25).
- The 0th-order source under $\alpha(\boldsymbol{s}) = 1/3$ is just the rigorous one.
- The sign of the (real part of the complex) boundary charge density is reversed in the anomalous dispersion regime, i.e.,

$$\left(\mathcal{R}e\left(\frac{\epsilon_1}{\epsilon_0}\right) + 2 \right) \left(\mathcal{R}e\left(\frac{\epsilon_1}{\epsilon_0}\right) - 1 \right) < -\left(\mathcal{I}m\left(\frac{\epsilon_1}{\epsilon_0}\right) \right)^2 . \tag{27}$$

If the inequality (27) is satisfied, $\mathcal{R}e(\epsilon_1/\epsilon_0)$ lies somewhere in the section $(-2, 1)$.

The scalar potential produced by the source (26) is

$$\phi(\boldsymbol{r}) = \phi^{(0)}(\boldsymbol{r}) \left(1 + f\left(\frac{a'}{r}\right) \frac{-1}{3} \frac{\epsilon_1 - \epsilon_0}{\frac{1}{3}\epsilon_1 + \frac{2}{3}\epsilon_0} \right) , \tag{28}$$

$$f\left(\frac{a'}{r}\right) \equiv \begin{cases} \left(\frac{a'}{r}\right)^3 & \text{for } \boldsymbol{r} \in \mathcal{V}_0 \\ 1 & \text{for } \boldsymbol{r} \in \mathcal{V}_1 \end{cases} .$$

Equation (28) coincides with the solution of (17a)–(17b) of the boundary value problem. Actually, one can easily check that the MBCs, i.e., (17b) and (19), are satisfied by (28).

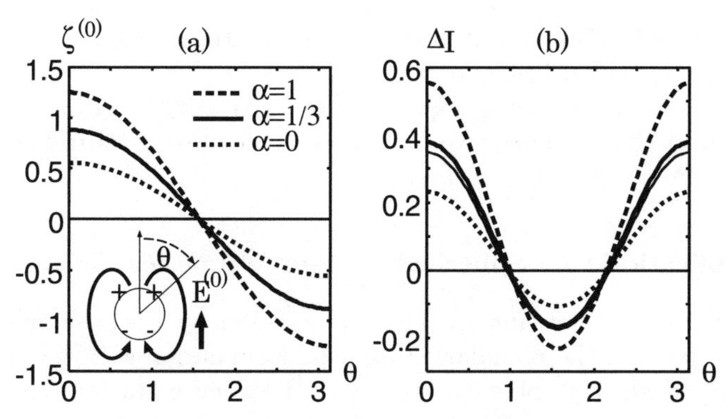

Fig. 6a,b. The analytical results for a spherical system under the NFC; we calculate based on our novel formulation with the scalar potential. We set $\epsilon_1/\epsilon_0 = 2.25$ and $\alpha = 0$, $1/3$ and 1. Note that the line with $\alpha = 1/3$ is just the rigorous one in the two graphs. (**a**) The 0th-order boundary source (the charge per unit area). (**b**) The electric-field intensity yielded by the 0th-order source; the observation points are on $a'/r = 1/1.5$ and the additional thin line is $\boldsymbol{E}^{(0)*} \cdot \Delta\boldsymbol{E} + \text{c.c.}$ under $\alpha = 1/3$, which is the main contribution to the intensity

Under a low-order approximation, the source and the field (intensity) depend on α, as is described in Fig. 6, where the line with $\alpha = 1/3$ coincides with the rigorous one. However, qualitative profiles for different αs are almost the same, e.g., the change of the sign of $\zeta^{(0)}$ and ΔI. This justifies the intuitive picture in Sect. 3.3 under a rough estimation of the source.

The sign of ΔI in Fig. 6b can be understood on the basis of (14). For example, at the positions around $\theta \sim 0$ [$\theta \sim \pi/2$], $\boldsymbol{E}^{(0)}$ and $\Delta\boldsymbol{E}$ are parallel [antiparallel], so that $\Delta I > 0$ [$\Delta I < 0$].

The above analytical calculation in the spherical system gives two hints, which are useful to treat a low symmetric system. One is that the arbitrariness in the boundary source may be used to improve the convergence in a numerical calculation for a low symmetric system. Actually, the condition for convergence of the above perturbation series in (25) leads to the next inequality,

$$\left|\alpha - \frac{1}{3}\right| < \left|\alpha - \frac{-\epsilon_0}{\epsilon_1 - \epsilon_0}\right| . \tag{29}$$

For a given complex-valued ϵ_1/ϵ_0 one can easily find such an α that satisfies (29). Because the l.h.s. (r.h.s.) is the distance between α and $1/3$ $[-\epsilon_0/(\epsilon_1 - \epsilon_0)]$ on the Gauss plane. Even in the low symmetric case it is expected that the convergence in a numerical calculation is improved by adjusting the arbitrariness.

The other is that the 0th-order source under $\alpha(\boldsymbol{s}) = 1/3$ may be used to obtain an intuitive picture of the electric near field in a low symmetric system, see Sect. 3.3. Actually in the spherical symmetric case, the 0th-order source under $\alpha(\boldsymbol{s}) = 1/3$ is just the rigorous one, see (26). Even if the shape of the matter is not spherical but isotropic enough, the 0th-order source under $\alpha(\boldsymbol{s}) = 1/3$ is expected to be good.

Although we do not check the case that α is a function on the boundary, we expect that an adequate function $\alpha(\boldsymbol{s})$ is also useful to improve the convergence in a numerical calculation.

3.9 Application to a Spherical System: Numerical Treatment

In this section, we show numerical calculations in a spherical symmetric system on the basis of the boundary scattering formulation with the scalar potential. We consider a spherical matter modeled by a stack of small cubes and perform numerical calculations under various values of the arbitrary parameter in the boundary charge density; we set $\alpha(\boldsymbol{s}) = \alpha$ (const.) in each calculation.

There are two purposes for these calculations. One is to check the program code, comparing numerical results with the analytical one derived in Sect. 3.8. The other is to examine that the solution is independent of the arbitrariness. For the latter purpose, we set $\epsilon_1/\epsilon_0 = 1.5$, because the smallness of $|\epsilon_1/\epsilon_0|$ leads to convergence over a wide range of the arbitrary parameter.

The procedure of the numerical calculation is as follows.

1. A dielectric sphere $\mathcal{V}_{01}/\eta$ is considered as a set of the boundary elements, which are the outside squares of the stacked cubes. The side of the small cube is set to 1/20 of the diameter of the sphere. Please note the notation: "a" in this subsection stands for the diameter of the sphere, while "a'" in Sect. 3.8 stands for the radius of the sphere, i.e., $a = 2a'$.
2. The source (the charge per unit area) in each boundary element is assumed to be homogeneous and its value is estimated at the center of the boundary element. The 0th-order source $\zeta^{(0)}(\boldsymbol{s})$ is given by (23b) at each boundary element. The 1st-order source $\zeta^{(1)}(\boldsymbol{s})$ at the center position of a selected boundary element is calculated by (23a); we replace $\zeta(\boldsymbol{s}')$ in the integral to $\zeta^{(0)}(\boldsymbol{s}')$ and regard the integral over $\mathcal{V}_{01}/\eta$ as the summation of each analytical integral over a single boundary element. After scanning $\boldsymbol{s}$ over all the center points of the boundary elements, one obtains a set of $\zeta^{(1)}(\boldsymbol{s})$.
3. In the same way, the 2nd- and higher-order source is obtained by solving (23a) iteratively. The convergence in every iteration is monitored by the standard deviation defined as

$$\text{s.d.} \equiv \left[\frac{\int_{\mathcal{V}_{01}/\eta} \mathrm{d}^2 s |\epsilon_0 \partial_n \phi^{(n)}(\boldsymbol{s}_0) - \epsilon_1 \partial_n \phi^{(n)}(\boldsymbol{s}_1)|^2}{|\epsilon_0 \boldsymbol{E}^{(0)}|^2 \int_{\mathcal{V}_{01}/\eta} \mathrm{d}^2 s} \right]^{1/2} . \tag{30}$$

   This standard deviation becomes zero, if the MBC (17b) is satisfied.
4. $\boldsymbol{E}(\boldsymbol{r})$ and $\Delta I(\boldsymbol{r})$ are calculated from the converged boundary source.

The cases under $\alpha = 0$, $1/3$, $2/3$, and 1 are examined. In the calculation under each α, the source is converged to the common one, i.e., the standard deviation in each case decreases monotonically as the number of the iteration increases and reaches the value of 1×10^{-6} or less for the 10th-order source. Therefore, the numerical solutions under various αs satisfy the MBCs and lead to the same profile of the field intensity. Furthermore, the common intensity profile based on the above numerical calculations coincides with that of the analytical one within the accuracy of the standard deviation. In Fig. 7, there are the intensity profiles only under $\alpha = 0$ and $\alpha = 1$ but those under the other αs are the same.

As a result, the numerical calculations based on the boundary scattering formulation with the scalar potential have been performed successfully.

3.10 Application to a Low Symmetric System

The next application of the boundary scattering formulation is a numerical calculation for a low symmetric system in order to emphasize the wide applicability of our formulation. The system is the same as that in [12][2]. A piece of matter is a dielectric ($\epsilon_1/\epsilon_0 = 2.25$) with the shape of the letter 'F'; the size of the longest side is 50 nm and it is placed in the vacuum, see Fig. 8a. It is exposed to incident light with the wavelength 633 nm and the observation point is assumed to be on the plane 5 nm below the piece of matter. This system satisfies the NFC and our formulation with the scalar potential is applicable. Note that evidence for the quasistatic picture has already been apparent in [12], although its authors did not mention; Figs. 1c–d of [12] are $\boldsymbol{k}$-independent.

The boundary of the piece of matter is divided into small pieces of squares of size 2.5 nm $\times$ 2.5 nm. The procedure of the numerical calculation is the same as that in Sect. 3.9. The labor of the coding program and the size of the calculation is much smaller than those in the full-retarded formulation, because only the two-dimensional source appears in our nonretarded formulation, while a three-dimensional source is needed in a full-retarded one, e.g., the formulation developed in Sects. 4 and 5.

The convergence in every iteration is monitored by the standard deviation (30) concerning the MBC (17b). The standard deviation decreases monotonically as the order of approximation increases and easily becomes the order of 10^{-4}; the convergence of the calculation is confirmed. The calculations under different αs are converged to the same contour map of the intensity as Fig. 8c, within the error of the s.d. Comparing our result with Fig. 1a of [12], one will find good coincidence, although we ignore all the retardation effects.

Let us explain for the contour map intuitively following the idea in Sect. 3.3. Place the 0th-order boundary source estimated by (23b) under

[2] Note that their intensity is just the square of the electric field, while ours is nondimensional and the background intensity is eliminated.

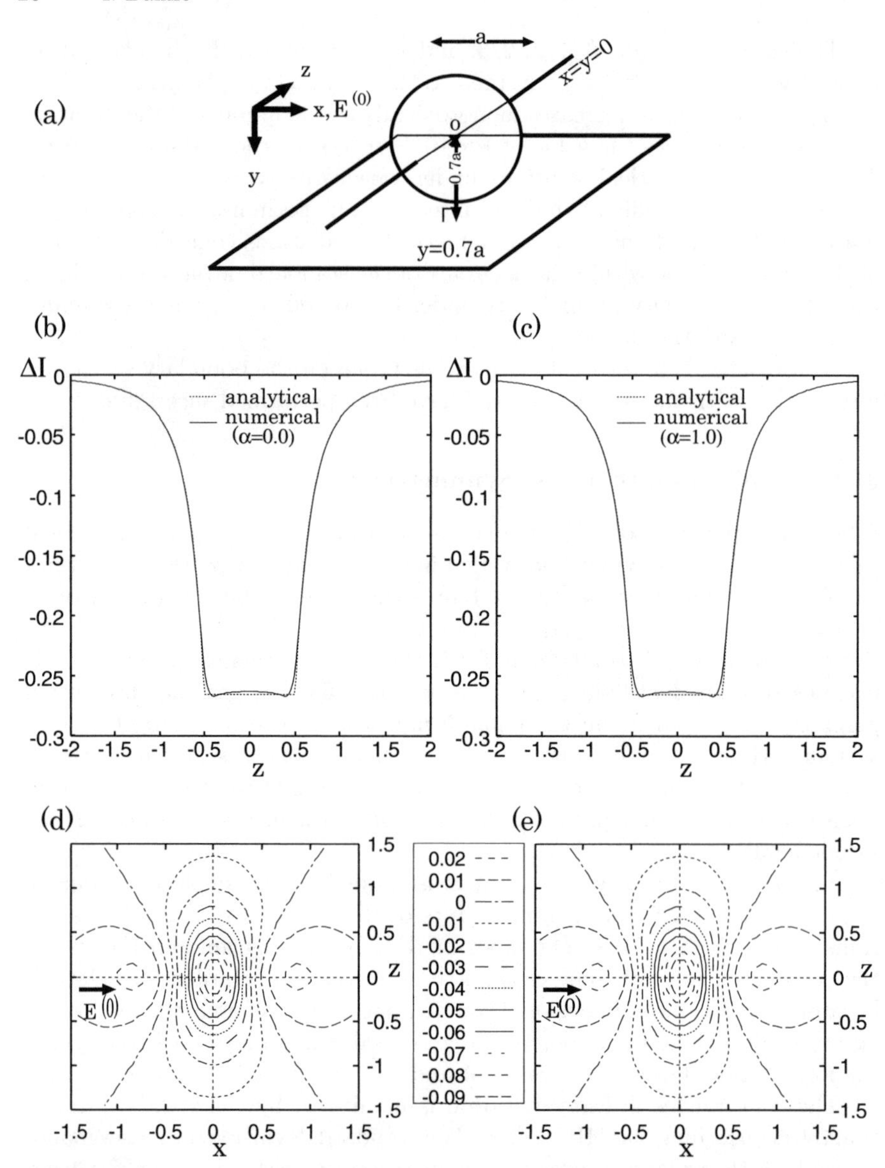

Fig. 7a-e. The electric near field of a dielectric sphere under the NFC; the numerical calculations are performed based on our novel formulation. (**a**) The definition of the coordinates. We set $a = 1$ (the diameter) and $\epsilon_1/\epsilon_0 = 1.5$. (**b**) The intensity profile along the line $x = y = 0$ yielded by the 10th-order source under $\alpha = 0.0$. The *dotted line* is that for the analytical solution. (**c**) The same as (**b**) under $\alpha = 1.0$. (**d**) The contour map of the electric intensity on the plane $y = 0.7a$ yielded by the 10th-order source under $\alpha = 0.0$. (**e**) The same as (**d**) under $\alpha = 1.0$

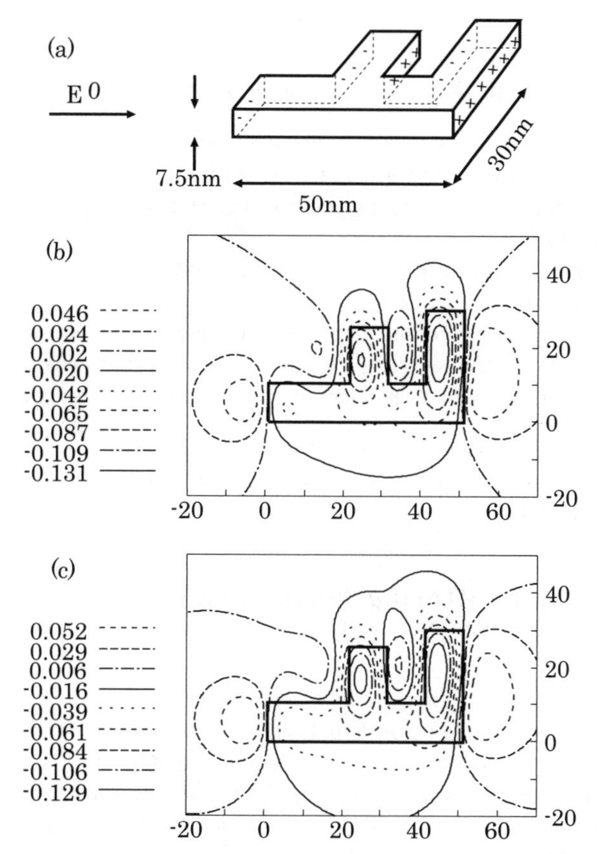

Fig. 8a-c. The electric near field of an 'F'-shaped dielectric under the NFC; the numerical calculations are performed based on our novel formulation. (**a**) The shape of the matter and the distribution of the 0th-order boundary source. (**b**) The contour map of the intensity produced by the 0th boundary source under $\alpha = 1/3$; the intensity is observed on the plane below the matter by 5 nm. (**c**) The contour map of the intensity produced by the 7th boundary source under $\alpha = 1/3$. Standard deviation is $\sim 1 \times 10^{-4}$

$\alpha = 1/3$ (Fig. 8a), imagine the electric flux and $\Delta \boldsymbol{E}$ and consider the inner product between $\Delta \boldsymbol{E}$ and $\boldsymbol{E}^{(0)}$. Then, the intensity profile in Fig. 8, including the change of the sign, is understood in the same way as is used for Fig. 6.

Among the intensity profile based on the 0th-order source under $\alpha = 0$, 1/3, 1/2, and 1, that under $\alpha = 1/3$ (see Fig. 8b) is most similar to the well-converged one, i.e., that under the 7th-order source, see Fig. 8c; this advantage to use $\alpha = 1/3$ has been pointed out in Sect. 3.8.

3.11 Summary

The essential points in this section are as follows:

- Under the NFC, the quasistatic picture holds; it can be understood intuitively on the basis of the static Coulomb law.
- The scalar potential is the minimum degree of freedom in the EM field under the NFC.
- The boundary value problem in a system with a steep interface can be replaced by a boundary scattering problem.
- The boundary charge density in this formulation possesses arbitrariness originating from the source's profile in the interface region. This arbitrariness does not affect the electric field outside the interface region.

In short, the boundary scattering formulation with the scalar potential gives a clear physical picture and a concise procedure to calculate the electric near field under NFC.

4 Boundary Scattering Formulation with Dual EM Potential

In order to discuss the EM near field with the retardation effect, we would like to reconstruct the electromagnetism with matter in terms of the dual EM potential. In this section, we introduce the *dual EM potential* as the minimum degree of freedom of the EM field under the assumption that the magnetic response of the matter is negligible [16]. Furthermore, a boundary value problem at a steep interface is replaced by *a boundary scattering problem*, i.e., a scattering problem with an adequate boundary source. This replacement is analogous to that in the scalar potential version discussed in Sect. 3.6.

The dual EM potential can be used in any classical EM problem with matter that possesses a negligible magnetic response. The dual EM potential together with the boundary scattering formulation is especially appropriate for the EM near field under the existence of the retardation effect; an application to such a EM near-field problem will appear in Sect. 5.

There is an additional merit in using dual/ordinary EM potential for EM problem with matter. The wave equation for the ordinary/dual EM potential in a stationary system is a vector Helmholtz equation; this wave equation for the EM field is equivalent to that for the de Broglie field with spin = 1 in quantum mechanics. Therefore, in order to understand the behavior of a EM field, we can make use of well-established concepts in quantum mechanics, e.g., the tunneling effect, the bound state, the spin-orbit interaction, and so on. An example of this idea will appear in Sect. 5. A further discussion based on this analogy will be given elsewhere in the future.

Table 2. Dual transformation in Maxwell's equations and equations for matter

Electric quantity	Magnetic quantity		Magnetic quantity	Electric quantity
$\boldsymbol{D}$	$\boldsymbol{B}$	$\Longrightarrow$	$\boldsymbol{B}$	$\boldsymbol{D}$
$\boldsymbol{E}$	$\boldsymbol{H}$	$\Longrightarrow$	$-\boldsymbol{H}$	$-\boldsymbol{E}$
$\boldsymbol{P}$	$\boldsymbol{M}$	$\Longrightarrow$	$\boldsymbol{M}$	$\boldsymbol{P}$
ϵ_0	μ_0	$\Longrightarrow$	$-\mu_0$	$-\epsilon_0$

4.1 Dual EM Potential as Minimum Degree of Freedom

In the ordinary EM problem with matter, there is only induced electric charge density and current density i.e., no external electric charge density nor current density. Maxwell's equations for this case are

$$\begin{aligned} \nabla \cdot \boldsymbol{B}(\boldsymbol{r},t) = 0,\ \nabla \times \boldsymbol{E}(\boldsymbol{r},t) + \partial_t \boldsymbol{B}(\boldsymbol{r},t) = \boldsymbol{0}\ , \\ \nabla \cdot \boldsymbol{D}(\boldsymbol{r},t) = 0,\ \nabla \times \boldsymbol{H}(\boldsymbol{r},t) - \partial_t \boldsymbol{D}(\boldsymbol{r},t) = \boldsymbol{0}\ , \end{aligned} \tag{31}$$

and the equations for matter are

$$\begin{aligned} \boldsymbol{B}(\boldsymbol{r},t) &= \mu_0 \boldsymbol{H}(\boldsymbol{r},t) + \boldsymbol{M}(\boldsymbol{r},t)\ , \\ \boldsymbol{D}(\boldsymbol{r},t) &= \epsilon_0 \boldsymbol{E}(\boldsymbol{r},t) + \boldsymbol{P}(\boldsymbol{r},t)\ , \end{aligned} \tag{32}$$

where the notations are as follow: $\boldsymbol{E}$ is electric field, $\boldsymbol{B}$ is magnetic flux density, $\boldsymbol{H}$ is magnetic field, $\boldsymbol{D}$ is electric flux density (or displacement vector field), $\boldsymbol{P}$ is polarization of the matter and $\boldsymbol{M}$ is magnetization of the matter.

Now note that there is dual symmetry in (31) and (32). Actually (31) and (32) are invariant under the dual transformation, i.e., the mutual exchange between the electric quantities and the magnetic ones in Table 2.

On the other hand, it is well known that the minimum degree of freedom of the EM field is neither $\boldsymbol{E}$ nor $\boldsymbol{B}$ but the EM potential $\boldsymbol{A}$ and ϕ. The ordinary EM potential is defined as $\boldsymbol{B} = \nabla \times \boldsymbol{A}$, $\boldsymbol{E} = -\partial_t \boldsymbol{A} - \nabla\phi$ and is especially convenient for electromagnetism with matter that possesses a certain magnetic response and no electric response, i.e., $\boldsymbol{M} \neq \boldsymbol{0}$ and $\boldsymbol{P} = \boldsymbol{0}$. Although such matter is not realistic, we discuss the EM field yielded by the source $\boldsymbol{M}$ as a guide to introduce the dual EM potential. Anyway, under the condition $\boldsymbol{M} \neq \boldsymbol{0}$ and $\boldsymbol{P} = \boldsymbol{0}$, the following wave equation for $\boldsymbol{A}$ in the radiation gauge is derived from (31) and (32).

$$\nabla \times \nabla \times \boldsymbol{A}(\boldsymbol{r},t) + \epsilon_0\mu_0 \partial_t^2 \boldsymbol{A}(\boldsymbol{r},t) = \nabla \times \boldsymbol{M}(\boldsymbol{r},t)\ , \tag{33a}$$

$$\nabla \cdot \boldsymbol{A}(\boldsymbol{r},t) = 0\ , \tag{33b}$$

$$\phi(\boldsymbol{r},t) = 0\ . \tag{33c}$$

In (33a)–(33c), the source of the field $\boldsymbol{A}$ is $\nabla \times \boldsymbol{M} = \mu_0 \times$(magnetizing electric current density); this transversal vector source yields the transversal field, i.e.

$\boldsymbol{A}$ in the radiation gauge. Note that the condition for the radiation gauge is satisfied in the whole space including the interface region.

The matter of interest to us is just dual to that discussed in the above paragraph, because the matter possesses a certain electric response and a negligible magnetic response, i.e., $\boldsymbol{P} \neq \boldsymbol{0}$ and $\boldsymbol{M} = \boldsymbol{0}$; in particular, this is the common case in the optical region. In connection with the duality, it is natural to introduce the *dual EM potential* defined as $\boldsymbol{D} = -\nabla \times \boldsymbol{C}$, $\boldsymbol{H} = -\partial_t \boldsymbol{C} - \nabla \chi$. Applying the dual transformation to (33a)–(33c), i.e., $(\boldsymbol{A}, \phi, \boldsymbol{M}) \Longrightarrow (-\boldsymbol{C}, -\chi, \boldsymbol{P})$, one obtains the wave equation followed by the dual EM potential,

$$\nabla \times \nabla \times \boldsymbol{C}(\boldsymbol{r},t) + \epsilon_0 \mu_0 \partial_t^2 \boldsymbol{C}(\boldsymbol{r},t) = -\nabla \times \boldsymbol{P}(\boldsymbol{r},t) \,, \tag{34a}$$

$$\nabla \cdot \boldsymbol{C}(\boldsymbol{r},t) = 0 \,, \tag{34b}$$

$$\chi(\boldsymbol{r},t) = 0 \,. \tag{34c}$$

The minimum degree of freedom of the EM field in our problem is the dual EM potential in the radiation gauge and its source is $\nabla \times \boldsymbol{P} = -\epsilon_0 \times$ (polarizing magnetic current density); this source is dual to $\nabla \times \boldsymbol{M} = \mu_0 \times$ (magnetizing electric current density) see (33a). Equations (34a)–(34c) are equivalent to Maxwell's equations under $\boldsymbol{M} = 0$. In fact, one can determine $\boldsymbol{D}, \boldsymbol{E}, \boldsymbol{B}$, and $\boldsymbol{H}$ from $\boldsymbol{C}$ using,

$$\boldsymbol{D}(\boldsymbol{r},t) = -\nabla \times \boldsymbol{C}(\boldsymbol{r},t) \,, \tag{35a}$$

$$\boldsymbol{E}(\boldsymbol{r},t) = \left(-\nabla \times \boldsymbol{C}(\boldsymbol{r},t) - \boldsymbol{P}(\boldsymbol{r},t)\right)/\epsilon_0 \,, \tag{35b}$$

$$\boldsymbol{B}(\boldsymbol{r},t) = \mu_0 \boldsymbol{H}(\boldsymbol{r},t) = -\mu_0 \partial_t \boldsymbol{C}(\boldsymbol{r},t) \,. \tag{35c}$$

4.2 Wave Equation for Dual Vector Potential

Suppose that the incident field oscillates with a constant angular frequency ω (or a constant wavenumber $k = \omega/c = \omega(\epsilon_0\mu_0)^{1/2}$) and that $\boldsymbol{P}$ is defined through a local and linear dielectric function $\epsilon(\boldsymbol{r})$ as,

$$\boldsymbol{P}(\boldsymbol{r},t) = (\epsilon(\boldsymbol{r}) - \epsilon_0)\boldsymbol{E}(\boldsymbol{r},t) = -\left(1 - \frac{\epsilon_0}{\epsilon(\boldsymbol{r})}\right) \nabla \times \boldsymbol{C}(\boldsymbol{r},t) \,. \tag{36}$$

$\epsilon(\boldsymbol{r})$ is assumed to be a smooth function for a time.

Omitting the common time dependence in $\boldsymbol{C}(\boldsymbol{r},t) = \boldsymbol{C}(\boldsymbol{r})\exp(-\mathrm{i}\omega t)$ and carrying out some calculations, (34a)–(34b) together with (36) lead to the following Helmholtz equation and the condition for the radiation gauge, i.e., (37a)–(37d); these equations are the starting point of our novel formulation.

$$\nabla\times\nabla\times\boldsymbol{C}(\boldsymbol{r})-k^2\boldsymbol{C}(\boldsymbol{r}) = -\hat{V}_s[\boldsymbol{C}](\boldsymbol{r})-\hat{V}_v[\boldsymbol{C}](\boldsymbol{r}) \text{ for } \boldsymbol{r}\in\mathcal{V}_0\cup\mathcal{V}_{01}\cup\mathcal{V}_1\ , \tag{37a}$$

$$\nabla\cdot\boldsymbol{C}(\boldsymbol{r}) = 0\ , \tag{37b}$$

$$\hat{V}_s[\boldsymbol{C}](\boldsymbol{r}) \equiv -\frac{\nabla\epsilon(\boldsymbol{r})}{\epsilon(\boldsymbol{r})}\times\nabla\times\boldsymbol{C}(\boldsymbol{r})\ , \tag{37c}$$

$$\hat{V}_v[\boldsymbol{C}](\boldsymbol{r}) \equiv -\Big(\frac{\epsilon(\boldsymbol{r})}{\epsilon_0}-1\Big)k^2\boldsymbol{C}(\boldsymbol{r})\ . \tag{37d}$$

The source is divided into two parts. One is (boundary magnetic current density)$\times(-\epsilon_0)$, namely $\hat{V}_s[\boldsymbol{C}]$, which is responsible for the MBCs, see Sect. 4.4. The other is (volume magnetic current density)$\times(-\epsilon_0)$, namely $\hat{V}_v[\boldsymbol{C}]$, which yields the retardation effect due to its k-dependence. Note that there is one more contribution to the retardation effect from the second term in the l.h.s. of (37a). In the following, we simply refer to $-\hat{V}_s[\boldsymbol{C}]$ $(-\hat{V}_v[\boldsymbol{C}])$ as the "boundary (volume) magnetic current density", ignoring the constant factor ϵ_0.

Equations (37a)–(37d) are equivalent to Maxwell's equations under $\boldsymbol{M}=0$ and are applicable both to the EM near field and to the EM far field. In particular, for a EM near-field problem under $ka \lesssim kr \lesssim 1$, we can consider both the boundary effect and the retardation effect on an equal footing; this application will appear in Sect. 5.

Physicists know from their experience that a description in terms of the minimum degree of freedom often brings a clear physical picture and/or convenience for calculation. In Sect. 3, we have already showed that the scalar potential is the minimum degree of freedom of the EM field under the NFC and that the description in terms of the scalar potential gives a clear physical picture and a concise procedure of numerical calculation. However, in all the EM problems with matter, the field $\boldsymbol{E}, \boldsymbol{B}$ and the volume source $\boldsymbol{P}, \boldsymbol{M}$ have been extensively used and the minimum degree of freedom, namely ordinary/dual EM potential, has never been used even in a simple problem like the reflection and/or the transmission phenomena at a plane interface. We speculate that this irrationality comes from a singularity that appears in the wave equation for the EM potential; that singularity is originated from the MBCs. Actually, if we introduce a steep interface through (15), $\hat{V}_s[\boldsymbol{C}]$ in (37a)–(37d) becomes a product of distributions and a mathematical difficulty arises; we have already met with such a singularity, namely the boundary charge density in Sect. 3. In order to use (37a)–(37d) instead of Maxwell's equations, we should control this singularity and this will be shown in Sects. 4.3 and 4.4.

4.3 Boundary Value Problem for Dual EM Potential

To overcome the difficulty to solve (37a)–(37d) in a system with a steep interface (15), one may replace the original problem based on (37a)–(37d) by a boundary value problem based on,

$$\nabla \times \nabla \times \boldsymbol{C}(\boldsymbol{r}) - k^2 \boldsymbol{C}(\boldsymbol{r}) = -\hat{V}_v[\boldsymbol{C}](\boldsymbol{r}) \qquad \text{for } \boldsymbol{r} \in \mathcal{V}_0 \cup \mathcal{V}_1 \,, \tag{38a}$$

$$\nabla \cdot \boldsymbol{C}(\boldsymbol{r}) = 0 \,, \tag{38b}$$

$$\hat{V}_v[\boldsymbol{C}](\boldsymbol{r}) \equiv -\theta(\boldsymbol{r} \in \mathcal{V}_1) \left(\frac{\epsilon_1}{\epsilon_0} - 1 \right) k^2 \boldsymbol{C}(\boldsymbol{r}) \,. \tag{38c}$$

$$\frac{1}{\epsilon_0} \boldsymbol{n}_s \times \nabla \times \boldsymbol{C}(\boldsymbol{s}_0) = \frac{1}{\epsilon_1} \boldsymbol{n}_s \times \nabla \times \boldsymbol{C}(\boldsymbol{s}_1) \qquad \text{for } \boldsymbol{s} \in \mathcal{V}_{01}/\eta \,, \tag{38d}$$

$$\boldsymbol{n}_s \cdot \boldsymbol{C}(\boldsymbol{s}_0) = \boldsymbol{n}_s \cdot \boldsymbol{C}(\boldsymbol{s}_1) \,, \tag{38e}$$

where $\boldsymbol{s}_0 = \boldsymbol{s} + 0\boldsymbol{n}_s$, $\boldsymbol{s}_1 = \boldsymbol{s} - 0\boldsymbol{n}_s$ (see Fig. 5) and $\theta(\boldsymbol{r} \in \mathcal{V}_1)$ is defined in (16). The boundary condition (38d) describes the continuity of the boundary-parallel component of the electric field and (38e) describes the continuity of the boundary-normal component of the magnetic flux density. These boundary conditions (38d)–(38e) are derived from (37a)–(37d) as follows. Keeping η finite and assuming that $\epsilon(\boldsymbol{r})$ in the interface region is smooth enough, (37a) is equivalent to [17],

$$\nabla \times \left(\frac{\epsilon_0}{\epsilon(\boldsymbol{r})} \nabla \times \boldsymbol{C}(\boldsymbol{r}) \right) = k^2 \boldsymbol{C}(\boldsymbol{r}) \,. \tag{39}$$

We obtain (38d), if we take the inner product of (39) with $\boldsymbol{m}_i$ ($i = 1$ or 2), integrate over the small area $\eta \otimes l_i$ in Fig. 5, apply Stokes' theorem and take the limit $\eta \to +0$. On the other hand, (38e) is obtained if one integrates the gauge condition (37b) over the small volume $\sigma \otimes \eta$, applies Gauss' theorem and takes the limit $\eta \to +0$. Ensure that an explicit formula for $\epsilon(\boldsymbol{r})$ in the interface region is not needed in the above derivation of (38d)–(38e). That is, the MBCs (38d)–(38e) are independent of a dielectric function in the interface region.

Outside the interface region, i.e., in $\mathcal{V}_0 \cup \mathcal{V}_1$, the solution of (37a)–(37d) under (15) is obtained by solving (38a)–(38e) of the boundary value problem. In the boundary value problem, the boundary conditions (38d)–(38e) contain sufficient information to construct the solution outside the interface region, therefore one does not require the source or the field in the interface region; see Sect. 8 for a detailed discussion starting from a given dielectric function in the interface region.

Note that there are two more MBCs;

$$\boldsymbol{n}_s \cdot \nabla \times \boldsymbol{C}(\boldsymbol{s}_0) = \boldsymbol{n}_s \cdot \nabla \times \boldsymbol{C}(\boldsymbol{s}_1) \,, \tag{40a}$$

$$\boldsymbol{n}_s \times \boldsymbol{C}(\boldsymbol{s}_0) = \boldsymbol{n}_s \times \boldsymbol{C}(\boldsymbol{s}_1) \,. \tag{40b}$$

Substituting (35a)–(35c) into (40a)–(40b) and (38d)–(38e), one obtains the familiar expressions for the MBCs in terms of $\boldsymbol{E}$, $\boldsymbol{D}$, $\boldsymbol{H}$, and $\boldsymbol{B}$. Equation (40a) describes the continuity of the boundary-normal component of the electric flux field. Equation (40b) describes the continuity of the boundary-parallel component of the magnetic field.

Equations (40a)–(40b) are trivial, because they are derived from an identity and an implicit assumption as follows; we keep η finite at first and afterward take the limit $\eta \to +0$. Equation (40a) is obtained by integrating the identity $\nabla \cdot \nabla \times \boldsymbol{C}(\boldsymbol{r}) = 0$ over the small volume $\sigma \otimes \eta$ in Fig. 5, applying Gauss' theorem and taking the limit $\eta \to +0$. Furthermore, (40b) comes from the implicit assumption that the singularity of $\boldsymbol{P}(\boldsymbol{r})$ is, at most, that of the step function. Under this assumption, the singularity of $\nabla \times \boldsymbol{C}(= -\boldsymbol{D})$ is that of the step function because of (36) and (15). Integrating $\nabla \times \boldsymbol{C}(\boldsymbol{r}) \sim$ (step function) over $\eta \otimes l_i$ $(i = 1, 2)$, applying Stokes' theorem and taking the limit $\eta \to +0$, one obtains (40b).

Therefore, we do not need the boundary conditions (40a)–(40b) in the calculation in terms of the dual EM potential.

It is troublesome to solve a boundary value problem in a low symmetric system. Even if one can obtain a solution fortunately, it is still difficult to give a clear physical picture. Furthermore, the effect of the boundary condition, namely boundary effect, and the retardation effect are treated in an unbalanced way, see Sect. 5.1. In the next subsection we will propose another way free from these difficulties in the boundary value problem.

4.4 Boundary Scattering Problem Equivalent to the Boundary Value Problem

To overcome the difficulty to solve (37a)–(37d) under (15), i.e., in a system with a steep interface, there is another way, the *boundary scattering formulation* with the dual EM potential. The difficulty in the original problem is that the boundary magnetic current density in (37a) is the product of distributions, which is not well defined in a general sense of mathematics. However, detailed analysis in Sect. 8 reveals that the boundary magnetic current density is a well-defined quantity and can be expressed in various ways.

Using one of the possible expressions for the boundary magnetic current density, (37a)–(37d) become (41a)–(41f), which are the elementary equations for the boundary scattering formulation with the dual EM potential.

$$\nabla \times \nabla \times \boldsymbol{C}(\boldsymbol{r}) - k^2 \boldsymbol{C}(\boldsymbol{r}) = -\hat{V}_s[\boldsymbol{C}](\boldsymbol{r}) - \hat{V}_v[\boldsymbol{C}](\boldsymbol{r}) \text{ for } \boldsymbol{r} \in \mathcal{V}_0 \cup \mathcal{V}_{01} \cup \mathcal{V}_1 \,, \tag{41a}$$

$$\nabla \cdot \boldsymbol{C}(\boldsymbol{r}) = 0 \,, \tag{41b}$$

$$\hat{V}_s[\boldsymbol{C}](\boldsymbol{r}) = \int_{\mathcal{V}_{01}/\eta} \mathrm{d}^2 s \delta^3(\boldsymbol{r} - \boldsymbol{s}) \frac{\epsilon_1 - \epsilon_0}{\epsilon(\boldsymbol{s})} \boldsymbol{n}_s \times \nabla \times \boldsymbol{C}(\boldsymbol{s}) \,, \tag{41c}$$

$$\epsilon(\boldsymbol{s}) \equiv \alpha(\boldsymbol{s})\epsilon_1 + (1 - \alpha(\boldsymbol{s}))\epsilon_0 \qquad \text{for } \boldsymbol{s} \in \mathcal{V}_{01}/\eta \,, \tag{41d}$$

$$\nabla \times \boldsymbol{C}(\boldsymbol{s}) \equiv \alpha(\boldsymbol{s}) \nabla \times \boldsymbol{C}(\boldsymbol{s}_1) + (1 - \alpha(\boldsymbol{s})) \nabla \times \boldsymbol{C}(\boldsymbol{s}_0) \,, \tag{41e}$$

$$\hat{V}_v[\boldsymbol{C}](\boldsymbol{r}) \equiv -\theta(\boldsymbol{r} \in \mathcal{V}_1) \left(\frac{\epsilon_1}{\epsilon_0} - 1 \right) k^2 \boldsymbol{C}(\boldsymbol{r}) \,. \tag{41f}$$

Equations (41d)–(41e) are merely the definitions of $\epsilon(\boldsymbol{s})$ and $\nabla \times \boldsymbol{C}(\boldsymbol{s})$, respectively; $\alpha(\boldsymbol{s})$ is an arbitrary smooth and complex-valued function on $\mathcal{V}_{01}/\eta$.

Here we only show that (41a)–(41f) of the boundary scattering problem leads to (38a)–(38e) of the boundary value problem. The derivation of (38a)–(38b) is trivial and the derivation of (38d)–(38e) is as follows.

Take the inner product of (41a) with $\boldsymbol{m}_i$ ($i = 1$ or 2), integrate over the infinitesimal volume $\sigma \otimes \eta$ ($\eta = +0$) in Fig. 5, apply Stokes' theorem to the l.h.s. over the infinitesimal area $l_i \otimes \eta$ and carry out the volume integral of the delta function in the r.h.s., one then obtains,

$$\boldsymbol{m}_i \cdot \boldsymbol{n}_s \times \nabla \times \boldsymbol{C}(\boldsymbol{s}_0) - \boldsymbol{m}_i \cdot \boldsymbol{n}_s \times \nabla \times \boldsymbol{C}(\boldsymbol{s}_1) = -\frac{\epsilon_1 - \epsilon_0}{\epsilon(\boldsymbol{s})} \boldsymbol{m}_i \cdot \boldsymbol{n}_s \times \nabla \times \boldsymbol{C}(\boldsymbol{s}) \,. \quad (42)$$

Equation (42) holds for $i = 1$ and 2, therefore, (42) without "$\boldsymbol{m}_i \cdot$" is true. Substitute (41d)–(41e) into (42) without "$\boldsymbol{m}_i \cdot$", then, one can obtain (38d) after some calculation. The arbitrary function $\alpha(\boldsymbol{s})$ disappears automatically and does not affect the field in $\mathcal{V}_0 \cup \mathcal{V}_1$. On the other hand, the MBC (38e) derived from (41b) in a similar way as discussed in Sect. 4.3.

In principle, the solution of (38a)–(38e) of the boundary value problem and that of (41a)–(41f) of the boundary scattering problem are equivalent in the domain $\mathcal{V}_0 \cup \mathcal{V}_1$. Because both the solutions in the regions $\mathcal{V}_0 \cup \mathcal{V}_1$ satisfy the same boundary conditions. However, (41a)–(41f) possesses considerable merits compared with (38a)–(38e). The first merit is that both the boundary effect and the retardation effect can be treated on an equal footing, while the two effects are treated in an unbalanced manner in (38a)–(38e) of the boundary value problem. The second is that the arbitrariness of the expression for the boundary source can be used to improve the convergence in a numerical calculation.

The arbitrariness in (41a)–(41f) comes from the degrees of freedom of the magnetic current density profile inside the interface region $\mathcal{V}_{01}$; not the detailed profile but the integrated magnetic current density over the width $\eta(= +0)$ determines the field in $\mathcal{V}_0 \cup \mathcal{V}_1$. This integrated quantity is analogous to the boundary charge density in the boundary scattering formulation with the scalar potential in Sect. 3.6. Another analogy is the multipole moment as is mentioned in Sect. 3.6. Therefore, it is reasonable that the arbitrariness of the boundary source appears. See Sect. 8 for a mathematical details for the boundary scattering formulation; the boundary magnetic current density is a well-defined quantity and is expressed with arbitrariness.

As a result, the MBCs can be built into the definition of the boundary magnetic current density and the boundary value problem based on (38a)–(38e) or the original problem based on (37a)–(37d) with (15) replaced by the boundary scattering problem based on (41a)–(41f).

In the next section, we will treat both the boundary effect and the retardation effect in a perturbative or an iterative method.

4.5 Integral Equation for Source and Perturbative Treatment of MBCs

Equations (41a)–(41f) are converted to the next integral equation.

$$\begin{aligned} \boldsymbol{C}(\boldsymbol{r}) = &\ \boldsymbol{C}^{(0)}(\boldsymbol{r}) \\ &+ \int_{\mathcal{V}_{01}/\eta} \mathrm{d}^2 s' \mathcal{G}^{(t)}(\boldsymbol{r}, \boldsymbol{s}') \cdot \frac{\epsilon_1 - \epsilon_0}{\epsilon(\boldsymbol{s}')} \boldsymbol{n}_s \times \nabla \times \boldsymbol{C}(\boldsymbol{s}') \\ &+ \int_{\mathcal{V}_1} \mathrm{d}^3 r' \mathcal{G}^{(t)}(\boldsymbol{r}, \boldsymbol{r}') \cdot \left(-\frac{\epsilon_1 - \epsilon_0}{\epsilon_0} k^2 \right) \boldsymbol{C}(\boldsymbol{r}') , \end{aligned} \tag{43}$$

where $\boldsymbol{C}^{(0)}(\boldsymbol{r})$ is the incident field and $\mathcal{G}_{ij}^{(t)}(\boldsymbol{r}, \boldsymbol{r}')$ is the transversal Green's function (tensor) for the vector Helmholtz equation; the explicit expression for this Green's function is given in Sect. 9.2.

Equation (43) leads to coupled integral equations for the volume source and the boundary source, i.e., $\boldsymbol{V}(\boldsymbol{r} \in \mathcal{V}_1) \equiv \boldsymbol{C}(\boldsymbol{r})$ and $\boldsymbol{S}(\boldsymbol{s} \in \mathcal{V}_{01}/\eta) \equiv \boldsymbol{n}_s \times \nabla \times \boldsymbol{C}(\boldsymbol{s})$; $\boldsymbol{V}$ and $\boldsymbol{S}$ determine the volume magnetic current density (41f), and the boundary magnetic current density (41c), respectively.

$$\begin{aligned} \boldsymbol{V}(\boldsymbol{r}) = &\ \boldsymbol{V}^{(0)}(\boldsymbol{r}) \\ &+ \int_{\mathcal{V}_{01}/\eta} \mathrm{d}^2 s' \mathcal{G}^{(t)}(\boldsymbol{r}, \boldsymbol{s}') \cdot \frac{\epsilon_1 - \epsilon_0}{\epsilon(\boldsymbol{s}')} \boldsymbol{S}(\boldsymbol{s}') \\ &+ \int_{\mathcal{V}_1} \mathrm{d}^3 r' \mathcal{G}^{(t)}(\boldsymbol{r}, \boldsymbol{r}') \cdot \left(-\frac{\epsilon_1 - \epsilon_0}{\epsilon_0} k^2 \right) \boldsymbol{V}(\boldsymbol{r}') \qquad \text{for} \quad \boldsymbol{r} \in \mathcal{V}_1 , \end{aligned} \tag{44a}$$

$$\boldsymbol{V}^{(0)}(\boldsymbol{r}) \equiv \boldsymbol{C}^{(0)}(\boldsymbol{r}) , \tag{44b}$$

$$\begin{aligned} \boldsymbol{S}(\boldsymbol{s}) = &\ \boldsymbol{S}^{(0)}(\boldsymbol{s}) \\ &+ \int_{\mathcal{V}_{01}/\eta} \mathrm{d}^2 s' \{ \alpha(\boldsymbol{s}) \boldsymbol{n}_s \times \nabla_{\boldsymbol{s}_1} \times \mathcal{G}^{(t)}(\boldsymbol{s}_1, \boldsymbol{s}') \\ &+ (1 - \alpha(\boldsymbol{s})) \boldsymbol{n}_s \times \nabla_{\boldsymbol{s}_0} \times \mathcal{G}^{(t)}(\boldsymbol{s}_0, \boldsymbol{s}') \} \cdot \frac{\epsilon_1 - \epsilon_0}{\epsilon(\boldsymbol{s}')} \boldsymbol{S}(\boldsymbol{s}') \\ &+ \int_{\mathcal{V}_1} \mathrm{d}^3 r' \{ \alpha(\boldsymbol{s}) \boldsymbol{n}_s \times \nabla_{\boldsymbol{s}_1} \times \mathcal{G}^{(t)}(\boldsymbol{s}_1, \boldsymbol{r}') \\ &+ (1 - \alpha(\boldsymbol{s})) \boldsymbol{n}_s \times \nabla_{\boldsymbol{s}_0} \times \mathcal{G}^{(t)}(\boldsymbol{s}_0, \boldsymbol{r}') \} \cdot \left(-\frac{\epsilon_1 - \epsilon_0}{\epsilon_0} k^2 \right) \boldsymbol{V}(\boldsymbol{r}') \\ &\qquad \text{for} \quad \boldsymbol{s} \in \mathcal{V}_{01}/\eta , \end{aligned} \tag{44c}$$

$$\boldsymbol{S}^{(0)}(\boldsymbol{s}) \equiv -\boldsymbol{n}_s \times \boldsymbol{D}^{(0)}(\boldsymbol{s}) , \tag{44d}$$

where $\epsilon(\boldsymbol{s}')$, and $\boldsymbol{S}(\boldsymbol{s}')$ appearing in above equations are estimated by (41d)–(41e). In a numerical calculation based on the boundary scattering formulation with the dual EM potential, the essential work is to solve (44a)–(44d). Once we obtain the sources $\boldsymbol{V}$ and $\boldsymbol{S}$, we can easily calculate the EM field using (43) together with (35a)–(35c).

The usual perturbative method can be applied to (44a)–(44d) and the solution satisfies the MBCs to a certain degree, according to the order of approximation.

Note that the rigorous solution $\boldsymbol{C}(\boldsymbol{r})$ must satisfy the next condition that is derived by taking the divergence of (41a),

$$\nabla \cdot \left(\hat{V}_s[\boldsymbol{C}](\boldsymbol{r}) + \hat{V}_v[\boldsymbol{C}](\boldsymbol{r})\right) = 0 \,. \tag{45}$$

Equation (45) implies the transversality of the total magnetic current density[3] Under a finite-order approximation, (45) is not satisfied, in particular, in the interface region $\mathcal{V}_{01}$. Therefore, under a finite-order approximation, the longitudinal component of the source possibly yields a longitudinal field so that the gauge condition (41b) and the MBC (38e) break down. However, in the procedure based on (44a)–(44d), the gauge condition (41b) and the MBC (38e) are satisfied under every order of approximation owing to $\mathcal{G}^{(t)}$ in (44a)–(44d); the longitudinal component of the source is filtered out by means of the contraction between the transversal vector Green's function $\mathcal{G}^{(t)}$ and the source. Therefore, in a practical numerical calculation, the condition (45) under every order of approximation is not important. Equation (45) is satisfied automatically, as long as the calculation is convergent enough.

After all, the boundary scattering formulation with the dual EM potential is free from MBCs and applicable both to the near-field problem and to the far-field problem. Furthermore, one can treat both the boundary effect and the retardation effect using a perturbative or an iterative method.

4.6 Summary

The essential points in this section are as follows:

- The dual EM potential is the minimum degree of freedom in the EM field with matter, of which the magnetic response is negligible.
- The boundary value problem in a system with a steep interface can be replaced by a boundary scattering problem. In this novel formulation,

[3] Equation (45) leads to the next two facts:
1. There is no motion of the magnetic pole; this is derived from (45) and the conservation law of the magnetic pole. (The absence of a single magnetic pole – we know from our experience – is a sufficient condition for (45).)
2. There is no monopole moment of the magnetic current;

$$\int \mathrm{d}^3 r \left(\hat{V}_s[\boldsymbol{C}](\boldsymbol{r}) + \hat{V}_v[\boldsymbol{C}](\boldsymbol{r})\right) = - \int \mathrm{d}^3 r \, \boldsymbol{r} \, \nabla \cdot \left(\hat{V}_s[\boldsymbol{C}](\boldsymbol{r}) + \hat{V}_v[\boldsymbol{C}](\boldsymbol{r})\right) = \boldsymbol{0}.$$

The 2nd part of the above equation is effective under the condition that the current is localized in a finite volume.

both the boundary effect and the retardation effect are treated on an equal footing.
- The boundary magnetic current density appearing in this formulation possesses arbitrariness originating from the source's profile in the interface region.

In short, the boundary scattering formulation with the dual EM potential is free from MBCs and is applicable to various EM problems.

5 Application of Boundary Scattering Formulation with Dual EM Potential to EM Near-Field Problem

In this section, we treat the EM near field under the condition $ka \lesssim kr \lesssim 1$ including the NFC. Under this condition, the boundary effect and the retardation effect are comparable and a balanced treatment of the two effects is needed. The boundary scattering formulation with the dual EM potential developed in Sect. 4 is appropriate not only to perform a numerical calculation but also to obtain an intuitive understanding of such a EM near field.

5.1 Boundary Effect and Retardation Effect

On the basis of the boundary scattering formulation with the dual EM potential, we can discuss both the boundary effect and the retardation effect on an equal footing. In order to emphasize a merit of this balanced treatment, let us estimate the magnitude of the EM field yielded by $\hat{V}_s$ and $\hat{V}_v$ in (41a)–(41f) under the next two conditions: the NFC and Rayleigh's far-field condition.

The electric field in the vacuum region $\mathcal{V}_0$ is equivalent to the electric flux (displacement vector) field and estimated by means of (43) using the 0th-order source.

$$\begin{aligned}
\boldsymbol{D}(\boldsymbol{r}) &= -\nabla \times \boldsymbol{C}(\boldsymbol{r}) \\
&\simeq \boldsymbol{D}^{(0)}(\boldsymbol{r}) \\
&\quad + \int_{\mathcal{V}_{01}/\eta} \mathrm{d}^2 s' \nabla \times \mathcal{G}^{(t)}(\boldsymbol{r}, \boldsymbol{s}') \cdot \frac{\epsilon_1 - \epsilon_0}{\epsilon(\boldsymbol{s}')} \boldsymbol{n}_{s'} \times \boldsymbol{D}^{(0)}(\boldsymbol{s}') \\
&\quad - \int_{\mathcal{V}_1} \mathrm{d}^3 r' \nabla \times \mathcal{G}^{(t)}(\boldsymbol{r}, \boldsymbol{r}') \cdot \left(-\frac{\epsilon_1 - \epsilon_0}{\epsilon_0} k^2 \right) \boldsymbol{C}^{(0)}(\boldsymbol{r}') \,. \qquad (46)
\end{aligned}$$

In the last expression in (46), the 1st term is the incident field, and is assumed to be $|\boldsymbol{D}^{(0)}| \sim \mathcal{O}(1)$ or equivalently $\boldsymbol{C}^{(0)} = \mathcal{O}(1/k)$; the 2nd term and the 3rd term are the fields yielded by the boundary magnetic current density and the volume magnetic current density, namely the boundary effect and the retardation effect, respectively.

In the last expression in (46), the factor of the boundary integral (without integrand) carries $\mathcal{O}(a^2)$ and that of the volume one carries $\mathcal{O}(a^3)$. We adopt

$\epsilon(\boldsymbol{s}') = \epsilon_0$ in the 2nd term for a rough estimation. The estimation for the above factors are common both under the NFC and the Rayleigh's far-field condition. The difference between the two cases comes from the factor of Green's function.

$$\begin{aligned} |\nabla \times \mathcal{G}^{(t)}(\boldsymbol{r} - \boldsymbol{r}'; k)| &= |\nabla \times \mathcal{G}(\boldsymbol{r} - \boldsymbol{r}'; k)| = |\nabla G(\boldsymbol{r} - \boldsymbol{r}'; k)| \\ &\simeq \left| \frac{1}{4\pi} \frac{\exp(ik|\boldsymbol{r} - \boldsymbol{r}'|)}{|\boldsymbol{r} - \boldsymbol{r}'|^2} (1 - ik|\boldsymbol{r} - \boldsymbol{r}'|) \right| . \end{aligned} \tag{47}$$

For details of Green's function in vector analysis, please see Sect. 9.

Under the NFC $ka \lesssim kr \ll 1$, the 1st term in the last expression in (47) is dominant and estimated as,

$$|\nabla \times \mathcal{G}^{(t)}(\boldsymbol{r} - \boldsymbol{r}'; k)| \sim \mathcal{O}\left(\frac{1}{r^2}\right) + \mathcal{O}\left(\frac{a}{r^3}\right) , \tag{48}$$

where we use a rough estimation of $|\boldsymbol{r} - \boldsymbol{r}'|$ under $r > r'$,

$$|\boldsymbol{r} - \boldsymbol{r}'| \simeq r - \hat{\boldsymbol{r}} \cdot \boldsymbol{r}' \sim \mathcal{O}(r) + \mathcal{O}(a) . \tag{49}$$

The term carrying $\mathcal{O}(1/r^2)$ in (48) couples with the monopole moment of the the magnetic current density; the monopole moment vanishes because of (45) (see footnote on p. 30). Therefore this term does not contribute to the field. After all, under the NFC, the contributions to the electric field from the boundary effect and that from the retardation effect are estimated as

$$\begin{aligned} &\text{the 2nd term in (46)} \sim \mathcal{O}\left(\frac{a^3}{r^3} \frac{\epsilon_1 - \epsilon_0}{\epsilon_0}\right) , \\ &\text{the 3rd term in (46)} \sim \mathcal{O}\left(ka \frac{a^3}{r^3} \frac{\epsilon_1 - \epsilon_0}{\epsilon_0}\right) . \end{aligned}$$

On the other hand, under Rayleigh's far-field condition $ka \ll 1 \ll kr$, the 2nd term in the last expression in (47) is dominant and estimated as

$$|\nabla \times \mathcal{G}^{(t)}(\boldsymbol{r} - \boldsymbol{r}'; k)| \sim \mathcal{O}\left(\frac{k}{r}\right) + \mathcal{O}\left(\frac{k}{r} ka\right) . \tag{50}$$

Ignoring the 1st term due to the absence of the monopole moment in the magnetic current density (see (45) and its footnote), one obtains

$$\begin{aligned} &\text{the 2nd term in (46)} \sim \mathcal{O}\Big((ka)^2 \frac{a}{r} \frac{\epsilon_1 - \epsilon_0}{\epsilon_0}\Big) , \\ &\text{the 3rd term in (46)} \sim \mathcal{O}\Big((ka)^3 \frac{a}{r} \frac{\epsilon_1 - \epsilon_0}{\epsilon_0}\Big) . \end{aligned}$$

The above results are summarized in Table 3. Under the NFC, the main contribution of the scattered field comes from the boundary effect and is independent of the wavenumber, i.e., the quasistatic picture holds as discussed

Table 3. The order estimation of scattered field amplitude under the near-field condition and Rayleigh's far-field condition

	Incident term		Boundary term		Volume term
Near-field condition	1	$\gtrsim$	$\left(\frac{a}{r}\right)^3 \frac{\epsilon_1 - \epsilon_0}{\epsilon_0}$	$\gg$	$ka \left(\frac{a}{r}\right)^3 \frac{\epsilon_1 - \epsilon_0}{\epsilon_0}$
Rayleigh's far-field condition	1	$\gg$	$(ka)^2 \frac{a}{r} \frac{\epsilon_1 - \epsilon_0}{\epsilon_0}$	$\gg$	$(ka)^3 \frac{a}{r} \frac{\epsilon_1 - \epsilon_0}{\epsilon_0}$

in Sect. 3.1. In fact, the electric field yielded by the boundary effect carries $(a/r)^3$ that reveals the magnitude of the electric field yielded by a static electric dipole moment; the relation of the static electric dipole moment and the boundary magnetic current density will be explained in Sect. 5.2. Under Rayleigh's condition, the boundary term carries the leading order and is dependent on "k". The far-field intensity is the square of the scattered field and carries $\mathcal{O}(k^4 a^6/r^2)$. This corresponds to the well-known expression for far-field intensity in the Rayleigh scattering problem. Now we use the formula for the far-field intensity. See Sect. 7 where we mention the difference between near-field intensity and far-field intensity.

Comparing the above two cases, one is convinced that the observation of Rayleigh's far field is k-dependent and bounded by the diffraction limit, while that of the near field under the NFC is free from the diffraction limit; such a difference is consistent with the result in Sects. 2.1–2.4.

5.2 Intuitive Picture Based on Dual Ampere Law under Near-field Condition

Under the NFC, the boundary effect is much larger than the retardation effect and a quasistatic picture holds, as discussed in Sect. 5.1. Concerning the quasistatic picture, we have discussed in Sect. 3.3 in the context of the boundary scattering formulation with the scalar potential; the Coulomb law governs the electric field. Now we show that the quasistatic picture is described by the dual Ampere law based on the boundary scattering formulation with the dual EM potential.

Ignoring all the retardation effects in (46), defining the scattered field by $\boldsymbol{\Delta D} \equiv \boldsymbol{D}(\boldsymbol{r}) - \boldsymbol{D}^{(0)}(\boldsymbol{r})$ and using $\nabla \cdot \boldsymbol{D}^{(0)}(\boldsymbol{r}) = 0$, one obtains the "dual Ampere law" [14–16], that is,

$$\nabla \times \boldsymbol{\Delta D}(\boldsymbol{r}) = \hat{V}_s[\boldsymbol{C}^{(0)}](\boldsymbol{r}) = -\frac{\epsilon_1 - \epsilon_0}{\epsilon(\boldsymbol{r})} \delta(\boldsymbol{r} \in \text{boundary}) \boldsymbol{n}_s \times \boldsymbol{D}^{(0)} , \quad (51)$$

where $\delta(\boldsymbol{r} \in \text{boundary})$ stands for the one-dimensional delta function in the direction of $\boldsymbol{n}_s$ and the value of $\epsilon(\boldsymbol{r})$ in the denominator is not well defined but a value between ϵ_0 and ϵ_1 is physically acceptable in a naive sense. It is enough to set $\epsilon(\boldsymbol{r}) = \epsilon_0$ for a rough treatment here. Under the NFC, the incident field $\boldsymbol{D}^{(0)}$ is regarded as constant over the whole system and the boundary source $\sim \boldsymbol{n}_s \times \boldsymbol{D}^{(0)}$ can be placed on each point of the boundary, e.g., the current on the boundaries of a dielectric cube in Fig. 9c.

Ensure that the ordinary Ampere law is $\nabla \times \boldsymbol{B}(\boldsymbol{r}) = \mu_0 \times$(electric current density) and your right hand is useful to understand the relation between the field and the source, see Fig. 9a. Dual to the ordinary Ampere law, (51) reveals that $-\epsilon_0 \times$(boundary magnetic current density) in the r.h.s. yields the scattered field $\Delta \boldsymbol{D}$ and your left hand is useful because of the negative sign in the r.h.s., see Fig. 9b. Then we can easily understand the electric flux of the scattered field yielded by the boundary magnetic current density.

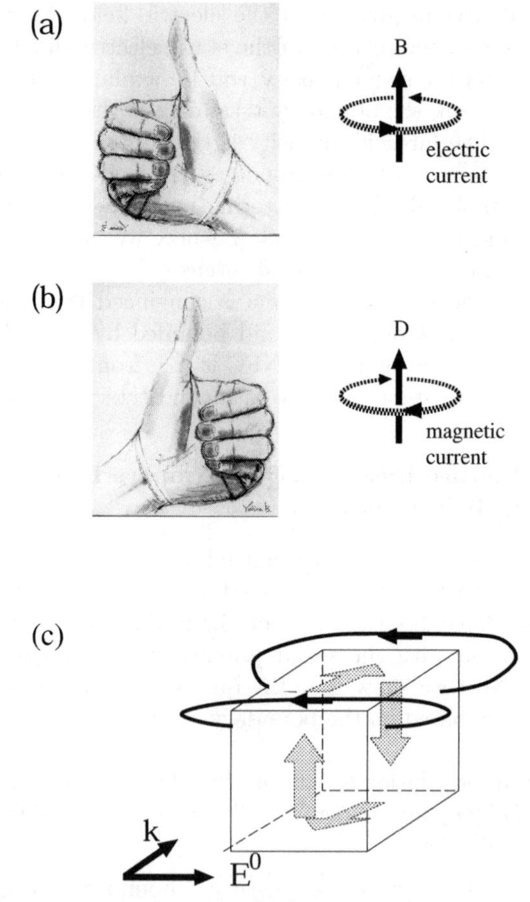

Fig. 9a-c. The relation between the source and the field (**a**) in the ordinary Ampere law and (**b**) in the dual Ampere law. (**c**) The electric flux yielded by the boundary magnetic current density under the NFC

Comparing Fig. 9c with Fig. 4b, it is found that the electric flux derived from the dual Ampere law is similar to that from the Coulomb law. The reason for this similarity is that the looped magnetic current is equivalent to a certain electric dipole moment, e.g., the pair of the boundary charge appeared on the opposite two boundaries in Fig. 4. This equivalence is dual to the well-known equivalence between a looped electric current and a magnetic dipole moment. Therefore, it is reasonable that the leading order of the scattered electric field under the NFC is estimated as $\mathcal{O}(a^3/r^3)$ in Sect. 5.1; this magnitude is just the same as that of the electric field yielded by a static dipole moment.

5.3 Application to a Spherical System: Numerical Treatment

Even in a spherical system, it is difficult to solve (41a)–(41f) of the boundary scattering problem analytically in the iterative method, while the corresponding boundary value problem can be solved analytically and is explained in familiar text books, e.g., [18]. In this section, we show numerical calculations in a spherical symmetric system on the basis of the boundary scattering formulation with the dual EM potential. We consider a sphere modeled by a stack of small cubes and perform numerical calculations under various values of the arbitrary parameter in the boundary source; we set $\alpha(\boldsymbol{s}) = \alpha$ (const.) in the whole calculations.

There are two purposes for these calculations. One is to check the program code by comparing numerical results with the analytical one in the text books. The other is to determine if the solution is independent of the arbitrariness $\alpha(\boldsymbol{s})$. For the latter purpose, we set $\epsilon_1/\epsilon_0 = 1.5$, because the smallness of $|\epsilon_1/\epsilon_0|$ leads to convergence over the wide range of the arbitrary parameter. In this section, the diameter of the sphere "a" together with "k" is fixed to $ka = 1$, so that the boundary effect and the retardation effect are comparable.

The procedure of the numerical calculation is as follows:

1. The body of the dielectric sphere $\mathcal{V}_1$ is considered as a set of volume elements that are small cubes, and the boundary of the sphere $\mathcal{V}_{01}/\eta$ is considered as a set of boundary elements that are the outside squares of the stacked cubes. The side of the small cube is set to 1/20 of the diameter of the sphere.
2. The volume (boundary) source in each volume (boundary) element is assumed to be homogeneous and its value is estimated at the center of the volume (boundary) element. The 0th-order volume (boundary) source is given by (44b) and (44d).
3. The coupled equations (44a)–(44d) are solved iteratively, in the analogous way explained in Sect. 3.9. The convergence in every iteration is monitored by the standard deviation defined as

$$\text{s.d.} \equiv \left[\frac{\int_{\mathcal{V}_{01}/\eta} \mathrm{d}^2 s |\boldsymbol{n}_s \times \boldsymbol{D}^{(n)}(\boldsymbol{s}_1)/\epsilon_1 - \boldsymbol{n}_s \times \boldsymbol{D}^{(n)}(\boldsymbol{s}_0)/\epsilon_0|^2}{|\boldsymbol{D}^{(0)}/\epsilon_0|^2 \int_{\mathcal{V}_{01}/\eta} \mathrm{d}^2 s} \right]^{1/2} . \quad (52)$$

This standard deviation becomes zero if the MBC (38d) is satisfied. We do not concern ourselves with the other MBCs; the MBC (38e) is automatically satisfied because $\mathcal{G}^{(t)}$ in (44a)–(44d) and the MBCs (40a)–(40b) are trivial, as discussed in Sect. 4.3.

4. $\boldsymbol{E}(\boldsymbol{r})$ and $\Delta I(\boldsymbol{r})$ are calculated from the converged boundary and volume sources.

The cases under $\alpha = 0$, $1/3$, $2/3$, and 1 are examined. In the calculation under each α, the source is converged to the common one, i.e., the standard deviation in each case decreases monotonically as the number of the iteration increases and reaches the value 1×10^{-4} or less for the 13th-order source. Therefore, the numerical solutions under various αs satisfy the MBC (38d) and lead to the same profile of the field intensity. Furthermore, the common intensity profile derived in the above numerical calculations coincides with that of the analytical one within the accuracy of the standard deviation. In Fig. 10, there are the intensity profiles only under $\alpha = 0$ and $\alpha = 1$ but those under the other αs are the same.

As a result, the numerical calculation based on the boundary scattering formulation with the dual EM potential have been performed successfully.

Although we do not check the case that α is a function on the boundary, we expect that an adequate function $\alpha(\boldsymbol{s})$ is also useful to improve the convergence in a numerical calculation.

5.4 Correction due to Retardation Effect

In the numerical result under $ka \lesssim kr \lesssim 1$, the intensity in the backside of the matter (i.e., $kz \lesssim -0.5$) is more negative than that in the frontside (i.e., $kz \gtrsim 0.5$), see Fig. 10. This asymmetric profile is different from the symmetric one in Fig. 7 under the NFC, i.e., $ka \lesssim kr \ll 1$, and should be attributed to the retardation effect.

In order to determine how the retardation effect works, we examine the EM near field of a dielectric cubes with $\epsilon_1/\epsilon_0 = 2.25$ in the three cases: $ka = 0.01, 0.10$, and 1.00, where "a" stands for the side length of the cube. The cubes are considered as a stack of small cubes, of which the side is set to 1/20 of that of the whole cube. The numerical calculation is based on the boundary scattering formulation with the dual EM potential and the procedure of the calculation is the same as that in Sect. 5.3. The results are shown in Fig. 11.

Let us put the difference of "ka" down to the difference of "k" keeping "a" constant. In this point of view, Fig. 11a and b are regarded as the same profile, independent of "k". This k-independence is understood as a characteristic of the quasistatic picture because the NFC is satisfied in the systems of Fig. 11a and b. In other words, the wavelength is so large that these systems cannot feel "k", see Fig. 3.

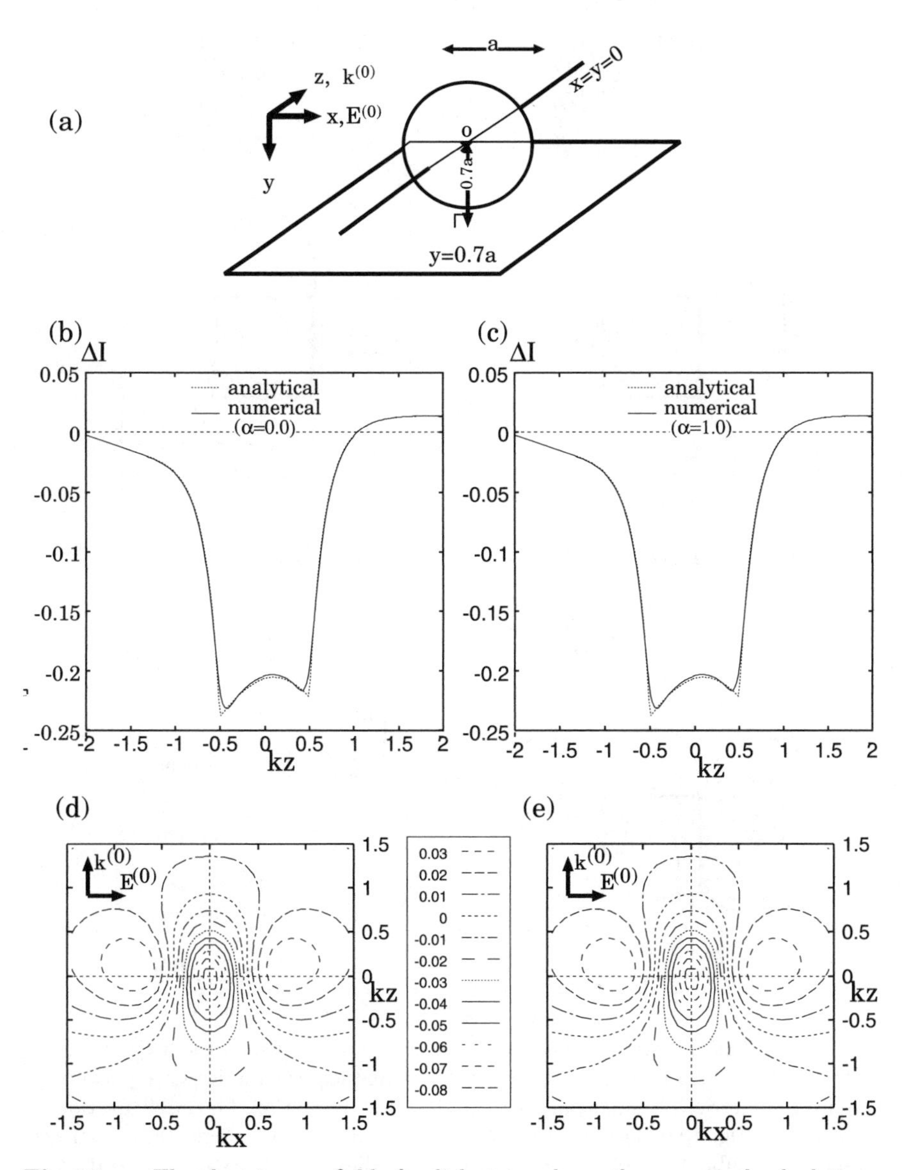

Fig. 10a-e. The electric near field of a dielectric sphere; the numerical calculations are performed based on our novel formulation with the dual EM potential. (**a**) Definition of coordinates. We set $ka = 1$ and $\epsilon_1/\epsilon_0 = 1.5$. (**b**) Intensity profile along the line $x = y = 0$ yielded by the 10th-order source under $\alpha = 0.0$. The *dotted line* is that for the analytical solution, which are seen in ordinary text books, e.g., [18]. (**c**) The same as (**b**) under $\alpha = 1.0$. (**d**) The contour map of the intensity on the plane, $y = 0.7a$ yielded by the 10th-order source under $\alpha = 0.0$. (**e**) The same as (**d**) under $\alpha = 1.0$

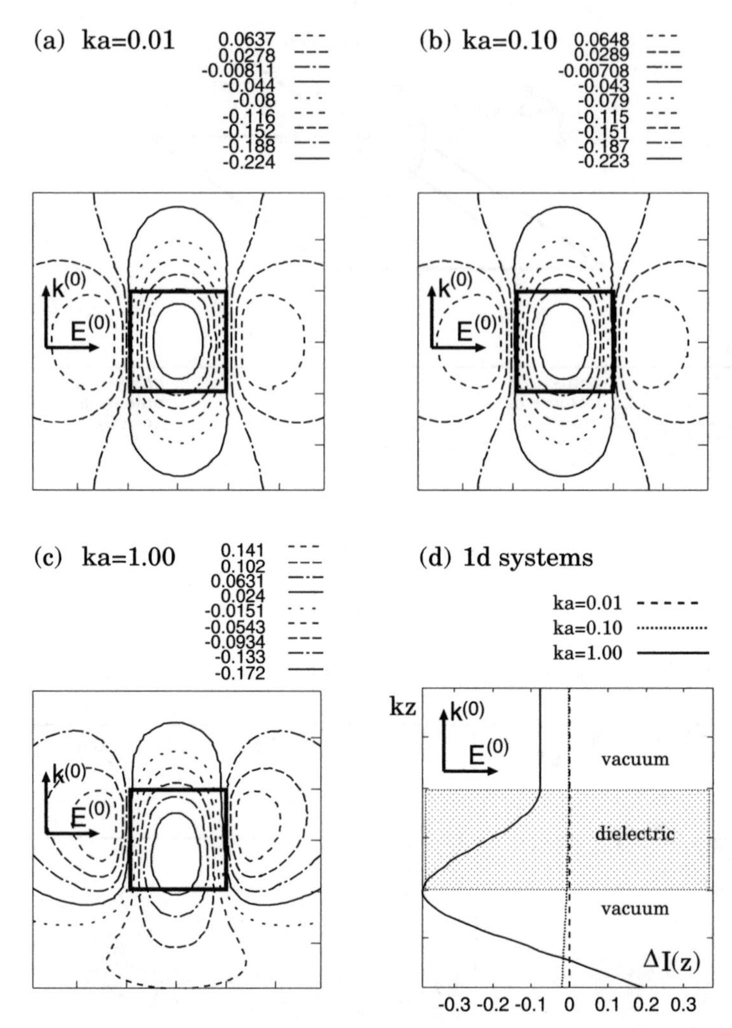

Fig. 11a-d. The electric near field of a dielectric cube, of which the dielectric constant is $\epsilon_1/\epsilon_0 = 2.25$ and side length is a; the numerical calculations are performed based on our novel formulation with the dual EM potential. The configuration is the same as Fig. 10a; the observation plane is located at the hight $0.7a$ from the center of the cube. (**a**), (**b**), (**c**) The intensity profiles for $ka = 0.01$, $ka = 0.10$, and $ka = 1.00$, respectively. (**d**) The intensity profile in the one-dimensional problem with normal incidence, i.e., s-polarization incidence

Equivalently we can put the difference of "ka" down to that of "a", keeping "k" constant. In this point of view, the commonness in Fig. 11a and b is considered as an invariant profile under the scale transformation concerning the length. This comes from the fact that the wave equation (37a)–(37d) or (41a)–(41f) in the limit of $k \to 0$ is invariant under the scale transformation concerning the spacial coordinates. This invariance is a property of the quasistatic picture and, of course, Poisson's equation, i.e., (12), (18) or (20a)–(20c), is also invariant under the scale transformation.

If the retardation effect is not negligible, "k" in the wave equation survives and this equation is not invariant under the scale transformation keeping "k" constant. Therefore, Fig. 11c is different from Fig. 11a and b. Analogous to Fig. 10, the intensity in the backside of the matter is more negative than that in the frontside.

In order to extract the retardation effect, let us consider a scattering problem with plane interfaces, namely a one-dimensional problem. Suppose that a dielectric occupies the region $-0.5 \le kz \le 0.5$ where the z direction is that of $\boldsymbol{k}$. Make sure that $\boldsymbol{k}$ of the incident field is normal to the interface, namely the s-polarized incident field. The whole field in this system possesses the same polarization vector as that of the incident field and can be expressed as,

$$\boldsymbol{C}(\boldsymbol{r}) = \frac{-\epsilon_0}{k^2} \nabla \times \hat{y} E(z) , \tag{53}$$

where $E(z) = \hat{y} \cdot \boldsymbol{E}(z)$ is the amplitude of the total electric field in question. Substituting (53) into (39) or (37a)–(37d), one obtains the next one-dimensional wave equation,

$$(-\partial_z^2 - k^2) E(z) = -\theta \left(-\frac{1}{2} \le kz \le \frac{1}{2} \right) \left(-\left(\frac{\epsilon_1}{\epsilon_0} - 1 \right) k^2 \right) E(z) . \tag{54}$$

In (54), there is no boundary source and the retardation effect survives. Solving (54) is equivalent to solving a quantum-well problem and the solution is obtained easily in connection with quantum mechanics. Note that the matter is dielectric, i.e., $\epsilon_1/\epsilon_0 - 1 > 0$ and the potential well corresponds to an attractive one in quantum mechanics. The result is Fig. 11d and it is found that the intensity in just the backside of the matter is more negative than that in the frontside. This is because the wave, propagating from the vacuum to the dielectric, feels the boundary $z = -0.5$ like the fixed end and the interference between the incident wave and the reflected one suppresses the amplitude just in the backside of the matter. In the frontside of the matter, however, there is no such destructive interference because there is no incident field from $z = +\infty$.

Now in our three-dimensional problem, both the boundary effect and the retardation effect contribute to the electric field. Under the condition $ka \lesssim kr \lesssim 1$, the field feels the boundary of the backside to some degree, because the width of this boundary is given by $ka = 1$ and is not negligible compared with the wavelength. Furthermore, the polarization vector is parallel to the boundary and it is similar to the s-polarization vector in the one-dimensional problem. Therefore, the contribution from the retardation effect is expected to be qualitatively the same as that of the one-dimensional problem. Now the field intensity formula (14) leads to

$$\begin{aligned}\Delta I(\boldsymbol{r}) &\simeq \frac{\boldsymbol{E}^{(0)*}(\boldsymbol{r}) \cdot \Delta \boldsymbol{E}(\boldsymbol{r}) + \text{c.c.}}{|\boldsymbol{E}^{(0)}(\boldsymbol{r})|^2} \\ &= \frac{\boldsymbol{E}^{(0)*}(\boldsymbol{r}) \cdot \Delta \boldsymbol{E}_{\text{surf}}(\boldsymbol{r}) + \text{c.c.} + \boldsymbol{E}^{(0)*}(\boldsymbol{r}) \cdot \Delta \boldsymbol{E}_{\text{vol}}(\boldsymbol{r}) + \text{c.c.}}{|\boldsymbol{E}^{(0)}(\boldsymbol{r})|^2}, \quad (55)\end{aligned}$$

where $\Delta \boldsymbol{E}_{\text{surf}}$ ($\boldsymbol{E}_{\text{vol}}$) is the scattered electric field that comes from the boundary (volume) integral in (43). $\Delta \boldsymbol{E}_{\text{surf}}$ contributes to the intensity in the same way both in the backside and in the frontside, at least, under the lowest-order approximation. However, $\Delta \boldsymbol{E}_{\text{vol}}$ contributes to the intensity in the backside more negatively than to that in the frontside even in the lowest-order approximation. In other words, $\Delta \boldsymbol{E}_{\text{vol}}$ in the backside is antiparallel to $\boldsymbol{E}^{(0)}$ and causes destructive interference. Summing up both the contributions to (55), it is confirmed that the asymmetry in the intensity profile Fig. 11c comes from the retardation effect.

At least, in some simple cases under $ka \lesssim kr \lesssim 1$, we expect that the EM near field is understood on the basis of the quasistatic picture with a certain correction due to the retardation effect, which is familiar in connection with quantum mechanics or wave mechanics.

5.5 Summary

The essential points in this section are as follows:

- The order of electric field under the NFC and that under Rayleigh's far-field condition are estimated on the basis of the boundary scattering formulation with the dual EM potential; the leading order in each case comes from the boundary effect and the field under NFC is "k"-independent, while Rayleigh's far field is "k"-dependent.
- Under the NFC, the quasistatic picture can be understood intuitively using the dual Ampere law; it is compatible with the picture in the context of the boundary scattering formulation with the scalar potential.
- Under the condition $ka \lesssim kr \lesssim 1$, a correction due to the retardation effect may be understood qualitatively in connection with the quantum mechanics.
- It is confirmed that the arbitrariness in the boundary scattering formulation does not affect the field outside the interface region.

In short, the boundary scattering formulation with the dual EM potential is useful to understand and to calculate the EM near field under the coexistence of the boundary effect and the retardation effect.

6 Summary and Remaining Problems

In short, what we have done are the following:

- Clear definitions of far field and near field are given.
- The boundary scattering formulations both with the scalar potential and with the dual EM potential are developed in order to treat the EM near field in a low symmetric system; both the formulations are free from the MBCs and enable a perturbative or an iterative treatment of the effect of the MBCs.
- A clear physical picture of EM near field on the basis of our formulations is presented.

The characteristics of the boundary scattering formulation with the scalar potential are the following:

- It is available under the NFC and is grounded upon the quasistatic picture. The minimum degree of freedom of the EM field under the NFC is the scalar potential.
- The MBCs are built into the boundary charge density; it possesses a certain arbitrariness, which never affects the physics. This formulation is free from the MBCs but equivalent to solving the corresponding boundary value problem.
- The boundary scattering problem can be solved by a perturbative or an iterative method. One can use the arbitrariness to improve the convergence in a numerical calculation.
- For the electric near field in the vicinity of a dielectric under the NFC, the lowest-order approximation of the perturbative treatment brings an intuitive picture based on the Coulomb law. This idea is effective even in a low symmetric system.
- It can be applied to a static-electric boundary value problem and a static-magnetic one.

The characteristics of the boundary scattering formulation with dual vector potential are the following:

- It is available, in principle, in all the regimes from near field to far field, because the wave equation for the dual vector potential in radiation gauge is equivalent to Maxwell's equations with matter, of which the magnetic response is negligible.
- The MBCs are built into the boundary magnetic current density; it possesses a certain arbitrariness, which never affects the physics. This formulation is free from the MBCs but equivalent to solving the corresponding boundary value problem.
- In this formulation, both the boundary effect and the retardation effect are treated on an equal footing. This balanced treatment is especially

appropriate to understand and to calculate the EM near field under $ka \lesssim kr \lesssim 1$; the two effects are comparable under this condition.
In the scheme of the boundary value problem, the two effects are treated in an unbalanced accuracy and a simple physical picture will never be obtained.

- Under the NFC, an intuitive picture based on the dual Ampere law holds. It is consistent with the picture based on the boundary scattering formulation with the scalar potential.
- Under the condition $ka \lesssim kr \lesssim 1$, the correction due to the retardation effect may be understood qualitatively in connection with the quantum mechanics.
- Under the condition $ka \lesssim kr \lesssim 1$, one can numerically calculate the EM near field by means of a perturbative or an iterative method. One can use the arbitrariness to improve the convergence.

Remaining problems are the following:

- Extension to treat optical effects of various types of matter, e.g., metal, magneto-optical matter and nonlinear matter, and so on. One may discuss the boundary optical effects within classical electromagnetism. In particular, those boundary effects are dominant in the near-field regime and should be considerably different from the well-known bulk or volume optical effects.
- Quantum theory on the basis of the dual EM potential.

7 Theoretical Formula for Intensity of Far Field, Near Field and Signal in NOM

Here we discuss theoretical formulas for far-field intensity and near-field intensity of an arbitrary scalar or vector field. Additional consideration is needed for a theoretical formula for the signal intensity in NOM [14,15].

7.1 Field Intensity for Far/Near Field

As a general starting point, the definition of the field intensity of an arbitrary scalar or vector field $X(\boldsymbol{r})$ is

$$\begin{aligned} \Delta I(\boldsymbol{r}) &= \frac{|X^{(0)}(\boldsymbol{r}) + \Delta X(\boldsymbol{r})|^2 - |X^{(0)}(\boldsymbol{r})|^2}{|X^{(0)}(\boldsymbol{r})|^2} \\ &= \frac{X^{(0)*}(\boldsymbol{r}) \cdot \Delta X(\boldsymbol{r}) + \text{c.c.} + |\Delta X(\boldsymbol{r})|^2}{|X^{(0)}(\boldsymbol{r})|^2} , \end{aligned} \tag{56}$$

where $|X^{(0)}(\boldsymbol{r})|^2$ in the denominator is introduced to make ΔI dimensionless and that in the numerator of the second part is to subtract the background intensity.

In the far-field observation ($kr \gg 1$), the observation point $\boldsymbol{r}$ does not belong to the coherent region of the incident field, i.e., $X^{(0)}(\boldsymbol{r}) = 0$. Therefore, (56) results in

$$\Delta I_{\mathrm{FF}}(\boldsymbol{r}) \simeq \frac{|\Delta X(\boldsymbol{r})|^2}{|X^{(0)}(\boldsymbol{r})|^2} \,. \tag{57}$$

We are familiar with this formula in the ordinary scattering theory.

In the near-field observation ($kr \lesssim 1$), the observation point $\boldsymbol{r}$ belongs to the coherent region of the incident field. Therefore, (56) under the assumption $|X^{(0)}| \gg |\Delta X|$ results in

$$\Delta I_{\mathrm{NF}}(\boldsymbol{r}) \simeq \frac{X^{(0)*}(\boldsymbol{r}) \cdot \Delta X(\boldsymbol{r}) + \mathrm{c.c.}}{|X^{(0)}(\boldsymbol{r})|^2} \,. \tag{58}$$

Equation (58) implies that the interference term is dominant in the near-field regime. The *interference effect* enables the negativeness of $\Delta I(\boldsymbol{r})$ in the near-field region, that is, the field intensity can be smaller than the background intensity. This fact is considerably different from the far-field intensity, which is positive definite.

In the above discussion, we implicitly assume that the multiple scattering effect between the sample and the probe (or detector) is negligible, therefore, we can express the field intensities without referring to a quantity of the probe. In the case of the far-field observation, this assumption is reasonable because the distance between the sample and the probe is very large. In the near-field observation, this assumption is justified if the size of the probe is small enough, i.e., $k \times$ (size of probe) $\ll 1$. In the recent experiments in NOM, this condition together with the near-field condition may be satisfied [8]; actually, the radius of curvature of the top of the probe tip is about 10 nm and is much smaller than the wavelength of the incident light, ~ 500 nm.

7.2 Theoretical Formula for the Signal Intensity in NOM

The signal intensity in NOM is considered as the field intensity of the transversal light; the light propagates in the optical fiber probe and possesses the polarization vector normal to the direction of the propagation. It is a rather simple assumption that the propagating field in the fiber is proportional to the near-field component normal to the direction of the fiber at the position of the probe tip. In this way, we may effectively take into account the *filtering effect* by the probe. The filtering effect has already been pointed out by others to explain the polarization dependence in an NOM image [19,20]. The filtered electric field is expressed by $-\boldsymbol{n}_{\mathrm{p}} \times \boldsymbol{n}_{\mathrm{p}} \times \boldsymbol{E}(\boldsymbol{r})$, where $\boldsymbol{n}_{\mathrm{p}}$ is the unit vector parallel to the direction of the fiber at the observation point in the near-field region. Then, a theoretical formula for the signal intensity in NOM is given by

$$\Delta I_{\mathrm{NOM}}(\boldsymbol{r}) \equiv \frac{\{\boldsymbol{n}_{\mathrm{p}} \times \boldsymbol{E}^{(0)*}(\boldsymbol{r})\} \cdot \{\boldsymbol{n}_{\mathrm{p}} \times \Delta \boldsymbol{E}(\boldsymbol{r})\} + \mathrm{c.c.} + |\boldsymbol{n}_p \times \Delta \boldsymbol{E}(\boldsymbol{r})|^2}{|\boldsymbol{E}^{(0)}(\boldsymbol{r})|^2} . \tag{59}$$

We again implicitly assume that the probe tip is so small that the multiple scattering effect is negligible, that is, we can express the signal intensity in NOM without using the properties of the probe except the filtering effect.

Now we can compare the signal intensity in NOM picked up by a small probe tip with the theoretical calculation based on the formula (59). See [14][4] for a qualitative comparison between the theoretical calculation and the experimental result in NOM.

8 Mathematical Basis of Boundary Scattering Formulation

Here we give a detailed discussion on the expressions for the boundary sources, i.e., the boundary charge density in (20a)–(20c) and the boundary magnetic current density in (41a)–(41f). We will make clear the following points:

- The boundary charge (magnetic current) density in (12) ((37a)–(37d)) is a well-defined quantity in the limit of the steep interface. It is the product of the delta function and the integrated charge (magnetic current) density over the infinitesimal width of the interface region.
- There are various expressions for the boundary source and that in (20a)–(20c) ((41a)–(41f)) is merely one of the possible expressions.
- The solution of (20a)–(20c) ((41a)–(41f)) in the boundary scattering problem is equivalent to that of (17a)–(17b) ((38a)–(38e)) in the corresponding boundary value problem outside the interface region.
- The arbitrariness of the expressions for the boundary source originates from the degrees of freedom of the source's distribution (or the dielectric function) in the interface region.

8.1 Boundary Charge Density and Boundary Condition

If the above points concerning the boundary charge density are proved in an arbitrary small volume in the vicinity of the interface region, it is true in the whole domain. Let us select a boundary element in the interface region, see Fig. 5, and restrict the domain of interest to the vicinity of the selected boundary element. We suppose the direction of x is that of $\boldsymbol{n}_s$ and the boundary element is located on the plane $x = 0$. At first, we keep the width of the interface region to be finite and give an arbitrary form of dielectric function there, then afterward we take the limit of the infinitesimal width.

[4] Note that the dual Ampere law is originally the Faraday law.

We define the dielectric function ϵ in the small volume of interest as follows,

$$\epsilon(x;\eta,[\xi]) \equiv \begin{cases} \epsilon_1 & \text{for} \quad x \leq -\dfrac{\eta}{2} \\ \xi\left(\dfrac{x}{\eta}\right) & \text{for} \quad -\dfrac{\eta}{2} < x < +\dfrac{\eta}{2} \\ \epsilon_0 & \text{for} \quad x \geq +\dfrac{\eta}{2} \end{cases} , \tag{60}$$

where $\xi(X)$, namely the smoothing function, is a complex-valued function defined in the real section $\{X|X \in \boldsymbol{R}, -1/2 < X < +1/2\}$ and satisfies the following conditions,

$$\xi(X) \in \mathcal{C}^1 , \quad \xi(X) \neq 0$$
$$\text{Re}(\xi(X)) \text{ and } \text{Im}(\xi(X)) \text{ are monotonic functions} , \tag{61a}$$
$$\lim\nolimits_{X\to-1/2}\xi(X) = \epsilon_1 , \quad \lim\nolimits_{X\to+1/2}\xi(X) = \epsilon_0 ,$$
$$\lim_{X\to-1/2}\frac{\mathrm{d}}{\mathrm{d}X}\xi(X) = \lim_{X\to+1/2}\frac{\mathrm{d}}{\mathrm{d}X}\xi(X) = 0 . \tag{61b}$$

For example, if one takes ξ as a function of degree three, the solution is

$$\xi(X) = 2(\epsilon_1 - \epsilon_0)X^3 + \frac{-3}{2}(\epsilon_1 - \epsilon_0)X + \frac{1}{2}(\epsilon_1 + \epsilon_0) .$$

Ensure that $\lim_{\eta\to+0}\epsilon(x;\eta,[\xi])$ corresponds to the dielectric function for the steep interface, see (15).

In the small volume of interest, (12) is reduced to the one-dimensional equation (62) in terms of $E(x) \equiv \boldsymbol{n}_s \cdot \boldsymbol{E}(x)$, i.e., the boundary-normal component of the electric field. Note that the boundary-parallel component of $\boldsymbol{E}$ in (12) is constant over the sufficiently small domain of interest.

Theorem for Boundary Charge Density

Consider the next equation,

$$\frac{\mathrm{d}}{\mathrm{d}x}E(x) = -\frac{\frac{\mathrm{d}}{\mathrm{d}x}\epsilon(x;\eta,[\xi])}{\epsilon(x;\eta,[\xi])}E(x) . \tag{62}$$

1. The solution of (62) under (60) is,

$$E(x) = E(x;\eta,[\xi]) = \frac{c_1}{\epsilon(x;\eta,[\xi])} , \tag{63}$$

where c_1 is independent of η and ξ, and satisfies the next relation,

$$\epsilon_0 E\left(+\frac{\eta}{2};\eta,[\xi]\right) = \epsilon_1 E\left(-\frac{\eta}{2};\eta,[\xi]\right) = c_1 . \tag{64}$$

2. Equations (62) and (60) in the limit of $\eta \to +0$ lead to the next equations of a boundary value problem.

$$\frac{\mathrm{d}}{\mathrm{d}x}E(x) = 0 \quad \text{for } x \neq 0\,, \tag{65a}$$

$$\epsilon_0 E_0 = \epsilon_1 E_1 = c_1\,, \tag{65b}$$

where $E_0 \equiv \lim_{\eta\to+0} E(+\frac{\eta}{2};\eta,[\xi])$, $E_1 \equiv \lim_{\eta\to+0} E(-\frac{\eta}{2};\eta,[\xi])$. c_1, E_0 and E_1 are independent of ξ.
In the three-dimensional problem, (17a)–(17b) in the boundary value problem are derived from (12) with the steep interface and the MBC (17b) is independent of a dielectric function in the interface region.
3. Equations (62) and (60) in the limit of $\eta \to +0$ lead to the next equation with a boundary charge density.

$$\frac{\mathrm{d}}{\mathrm{d}x}E(x) = c_1\left(\frac{1}{\epsilon_0} - \frac{1}{\epsilon_1}\right)\delta(x)$$
$$x \in \text{ the whole space of interest}\,, \tag{66}$$

where c_1 is given in (65b). $c_1(1/\epsilon_0 - 1/\epsilon_1)$ is the integrated charge density over the infinitesimal width of the interface region and independent of ξ. Make sure that c_1 is expressed in various ways in terms of E_0 and E_1, because (65b) carries two equalities among the three parameters, namely c_1, E_0, and E_1. Therefore, we can construct various boundary scattering problems based on (66), as is shown in the following; each boundary scattering problem is equivalent to the boundary value problem based on (65a)–(65b) in the region $x \neq 0$.
In the three-dimensional problem, the boundary charge density is a well-defined quantity, i.e., the product of the delta function and the integrated charge density over the infinitesimal width of the interface region; that integrated quantity is independent of the dielectric function in the interface region. Furthermore, we can express the boundary charge density in terms of the boundary value of the field and obtain a boundary scattering problem that is equivalent to the boundary value problem based on (17a)–(17b).
4. To obtain an expression for the boundary charge density, we may start from *any smoothing function* that satisfies (61a)–(61b). For example, we may take the next ξ

$$\xi(X) = \frac{\epsilon_0\epsilon_1}{\frac{1+2X}{2}\epsilon_0 + \frac{1-2X}{2}\epsilon_1}\,, \tag{67}$$

where $X \in \{X | X \in \boldsymbol{R}, -1/2 < X < +1/2\}$. Note that the domain is defined as an open section; outside the section, $\xi(X)$ should be an adequate function of X, which satisfies the conditions (61a)–(61b). Equation (66) together with (67) leads to

$$\frac{\mathrm{d}}{\mathrm{d}x}E(x) = \frac{\epsilon_1 - \epsilon_0}{\frac{1+2X^\bullet}{2}\epsilon_0 + \frac{1-2X^\bullet}{2}\epsilon_1}\left(\frac{1-2X^\bullet}{2}E_0 + \frac{1+2X^\bullet}{2}E_1\right)\delta(x)$$
$$x \in \text{ the whole space of interest}\,, \tag{68}$$

where $X^\bullet \in \{X | X \in \boldsymbol{R}, -1/2 < X < +1/2\}$ is the representative position in the interface region. Furthermore, the coefficient of the delta function in the r.h.s. of (68) is a function of $X^\bullet$ and can be extended to a function defined in a certain complex domain of $X^\bullet$ by means of "analytical continuation" [21]. In other words, although the boundary charge density is derived from $\xi(X)$, which is defined in the real and bounded domain of $X^\bullet$, the boundary charge density is effective beyond the initially assumed domain of $X^\bullet$.
If we compare (68) with (20a)–(20c), then we recognize that $\frac{1-2X^\bullet}{2}$ corresponds to the arbitrary complex-valued function $\alpha(\boldsymbol{s})$ and the r.h.s. of (68) corresponds to the boundary charge density in (20a)–(20c). Therefore, (68) justifies (20a)–(20c).

5. Furthermore, one may start from a different smoothing function ξ, the form of which is defined by means of some parameters. Then, one will trace the above procedure and derive a different expression for the boundary charge density, which includes $X^\bullet$ originating from the representative coordinate and the additional parameters introduced in the definition of ξ. Therefore, how to express the boundary charge density is rather arbitrary.
In short, the boundary charge density is expressed with a certain arbitrariness originated from the arbitrariness of the profile of the source (or the dielectric function) in the interface region.

Proof

1. Equation (62) in the interface region leads to,

$$\frac{\mathrm{d}}{\mathrm{d}X}\zeta(X) = -\frac{\frac{\mathrm{d}}{\mathrm{d}X}\xi(X)}{\xi(X)}\zeta(X), \quad \text{for} \quad -1/2 < X < +1/2\,, \tag{69}$$

where $X \equiv x/\eta$ and $\zeta(X) \equiv E(\eta X)$. One can easily solve (69) and can obtain

$$\xi(X)\zeta(X) = c_1\,, \tag{70}$$

where c_1 is an integral constant and is independent of η, because η does not appear in (69). Furthermore, c_1 is independent of ξ, because the magnitude of c_1 is merely the normalization factor of the field $\zeta(X)$, which follows the linear differential equation (69). Therefore, c_1 is not related to the function form of ξ. Equation (70) leads to (64) in the limit of $X \to \pm 1/2$. In short, the solution of (62) under (60) is (63).
2. The r.h.s. of (62) vanishes outside the interface region and leads to (65a) in the limit of $\eta \to +0$. On the other hand, the boundary condition (65b) is derived from (64) in the limit of $\eta \to +0$. Note that E_0 and E_1 are independent of ξ, because c_1 is independent of ξ, i.e., the MBC (65b) is free from the details of the dielectric function in the interface region.

3. Let us consider the next quantity,

$$g(x;\eta,[\xi]) \equiv \frac{1}{c_1\left(\frac{1}{\epsilon_0}-\frac{1}{\epsilon_1}\right)} \frac{-\frac{\mathrm{d}}{\mathrm{d}x}\epsilon(x;\eta,[\xi])}{\epsilon(x;\eta,[\xi])} E(x;\eta,[\xi])$$
$$= \frac{1}{\frac{1}{\epsilon_0}-\frac{1}{\epsilon_1}} \frac{\mathrm{d}}{\mathrm{d}x} \frac{1}{\epsilon(x;\eta,[\xi])}, \tag{71}$$

where we use (63) to obtain the last part. Taking the limit of $\eta \to +0$,

$$\lim_{\eta\to+0} g(x;\eta,[\xi]) = \lim_{\eta\to+0} \frac{1}{\eta} \left(\frac{\frac{1}{\epsilon_0}-\frac{1}{\epsilon_1}}{\eta}\right)^{-1} \frac{\mathrm{d}}{\mathrm{d}x}\frac{1}{\epsilon(x;\eta,[\xi])}$$
$$= \begin{cases} +\infty & \text{for } x=0 \\ 0 & \text{for } x \neq 0 \end{cases}. \tag{72}$$

To obtain the last expression, we use the fact that $\frac{\mathrm{d}}{\mathrm{d}x}(1/\epsilon(x;\eta,[\xi]))$ and $(\frac{1}{\epsilon_0}-\frac{1}{\epsilon_1})/\eta$ is identical in the interface region under the condition (61a), while $\frac{\mathrm{d}}{\mathrm{d}x}\frac{1}{\epsilon(x;\eta,[\xi])} = 0$ outside the interface region. On the other hand, an integral of $g(x;\eta,[\xi])$ times a sufficiently smooth function $f(x)$ becomes as follows in the limit of $\eta \to +0$,

$$\lim_{\eta\to+0} \int_{-\infty}^{+\infty} \mathrm{d}x f(x) g(x;\eta,[\xi])$$
$$= \lim_{\eta\to+0} \int_{-\eta/2}^{+\eta/2} \mathrm{d}x f(x) g(x;\eta,[\xi])$$
$$= \left(\frac{1}{\epsilon_0}-\frac{1}{\epsilon_1}\right)^{-1} \lim_{\eta\to+0} \int_{-1/2}^{+1/2} \mathrm{d}X f(\eta X) \frac{\mathrm{d}}{\mathrm{d}X}\frac{1}{\xi(X)}$$
$$= f(0)\left(\frac{1}{\epsilon_0}-\frac{1}{\epsilon_1}\right)^{-1} \int_{-1/2}^{+1/2} \mathrm{d}X \frac{\mathrm{d}}{\mathrm{d}X}\frac{1}{\xi(X)}$$
$$= f(0)\,. \tag{73}$$

From (72)–(73), $\lim_{\eta\to+0} g(x;\eta,[\xi])$ is just the delta function, therefore,

$$\lim_{\eta\to+0} \frac{-\frac{\mathrm{d}}{\mathrm{d}x}\epsilon(x;\eta,[\xi])}{\epsilon(x;\eta,[\xi])} E(x;\eta,[\xi]) = c_1\left(\frac{1}{\epsilon_0}-\frac{1}{\epsilon_1}\right)\delta(x)\,. \tag{74}$$

Equation (74) implies that the boundary charge density is a well-defined quantitiy and is independent of ξ in the limit of the steep interface, because c_1 is independent of ξ. Equations (62) and (74) lead to (66). Note that the factor $c_1(1/\epsilon_0 - 1/\epsilon_1)$ in the boundary charge density in (74) is the integrated charge density over the interface region, see (73) and (71). Furthermore, integrating (66) over the infinitesimal interface region including the selected boundary element, we obtain

$$E_0 - E_1 = c_1\left(\frac{1}{\epsilon_0} - \frac{1}{\epsilon_1}\right) = \epsilon_0 E_0\left(\frac{1}{\epsilon_0} - \frac{1}{\epsilon_1}\right) = E_0\left(1 - \frac{\epsilon_0}{\epsilon_1}\right) , \quad (75)$$

where we use one relation $c_1 = \epsilon_0 E_0$ in (65b) to derive the 3rd part. Equation (75) results in the other relation $\epsilon_0 E_0 = \epsilon_1 E_1$ in (65b). This means that (66) fully reproduces the MBC (65b) and is equivalent to (65a)–(65b) in the domain $x \neq 0$.

4. In the interface region, (67), (63) (or (70)), and (65b) lead to

$$\zeta(X) = \frac{1+2X}{2}\frac{c_1}{\epsilon_1} + \frac{1-2X}{2}\frac{c_0}{\epsilon_0} = \frac{1+2X}{2}E_1 + \frac{1-2X}{2}E_0 . \quad (76)$$

Therefore, setting X to the representative point $X^\bullet \in \{X | X \in \mathcal{R}, -1/2 < X < +1/2\}$, c_1 is expressed as

$$c_1 = \frac{\frac{1+2X^\bullet}{2}E_1 + \frac{1-2X^\bullet}{2}E_0}{\frac{1-2X^\bullet}{2}\frac{1}{\epsilon_0} + \frac{1+2X^\bullet}{2}\frac{1}{\epsilon_1}} . \quad (77)$$

Then, (66) and (77) result in (68).

5. Please trace the above procedure to obtain a different expression for the boundary charge density, starting from ξ as you like.

$\mathcal{QED}$.

8.2 Boundary Magnetic Current Density and Boundary Condition

Next we discuss the expression for the boundary magnetic current density in (41a)–(41f). As mentioned in the first part of Sect. 8.1, we concentrate on a sufficiently small volume including a selected boundary element in the interface region (Fig. 5). The notation for the coordinate and the definition of the dielectric function is the same as those in Sect. 8.1.

Then, ignoring all the retardation effect in (37a)–(37d) results in the one-dimensional equation (78) in terms of the boundary-parallel components of the displacement vector field $\boldsymbol{D}$. Ensure that the boundary-normal component of $\boldsymbol{D}$ is constant over the sufficiently small domain of interest.

Theorem for Boundary Magnetic Current Density

Consider the next equation.

$$\frac{\mathrm{d}}{\mathrm{d}x}D_i(x) = \frac{\frac{\mathrm{d}}{\mathrm{d}x}\epsilon(x)}{\epsilon(x)}D_i(x) \qquad i = 2, 3 , \quad (78)$$

where $D_2(x)$ and $D_3(x)$ stand for the y-component and z-component of $\boldsymbol{D}(x)$, respectively; these are the boundary-parallel components of $\boldsymbol{D}(x)$.

1. The solution of (78) under (60) is,

$$D_i(x) = D_i(x;\eta,[\xi]) = c_i\epsilon(x;\eta,[\xi])\ , \tag{79}$$

where c_i is independent of η and ξ, and satisfies the next relation,

$$\frac{D_i(+\frac{\eta}{2};\eta,[\xi])}{\epsilon_0} = \frac{D_i(-\frac{\eta}{2};\eta,[\xi])}{\epsilon_1} = c_i\ . \tag{80}$$

2. Equations (78) and (60) in the limit of $\eta \to +0$ lead to the next equations of a boundary value problem,

$$\frac{\mathrm{d}}{\mathrm{d}x}D_i(x) = 0 \qquad i = 2,3 \quad \text{for } x \neq 0\ , \tag{81a}$$

$$\frac{D_{i0}}{\epsilon_0} = \frac{D_{i1}}{\epsilon_1} = c_i\ , \tag{81b}$$

where $D_{i0} \equiv \lim_{\eta\to+0} D_i(+\frac{\eta}{2};\eta,[\xi])$, $D_{i1} \equiv \lim_{\eta\to+0} D_i(-\frac{\eta}{2};\eta,[\xi])$. c_i, D_{i0}, and D_{i1} are independent of ξ.
In the three-dimensional problem, we should revive the retardation effect but it does not affect the boundary condition. Therefore, the statement of this item implies that (38a)–(38e) in the boundary value problem are derived from (37a)–(37d) with the steep interface and that the MBC (38d) is independent of a dielectric function in the interface region. Note that the MBC (38e) comes from the gauge condition and is not related to the dielectric function by nature.
3. Equations (78) and (60) in the limit of $\eta \to +0$ lead to the next equation with a boundary magnetic current density

$$\frac{\mathrm{d}}{\mathrm{d}x}D_i(x) = -c_i\,(\epsilon_1 - \epsilon_0)\,\delta(x)$$
$$i = 2,3 \qquad x \in \text{the whole space of interest}\ , \tag{82}$$

where c_i is given in (81b). $c_i(\epsilon_1 - \epsilon_0)$ is the integrated magnetic current density over the infinitesimal width of the interface region and independent of ξ.
Ensure that c_i is expressed in various ways in terms of D_{i0}, and D_{i1}, because (81b) carries two equalities among the three parameters, namely c_i, D_{i0}, and D_{i1}. Therefore, we can construct various boundary scattering problems based on (82), as is shown in the following; each boundary scattering problem is equivalent to the boundary value problem based on (81a)–(81b) in the region $x \neq 0$.
In the three-dimensional problem, the boundary magnetic current density is a well-defined quantity, i.e., the product of the delta function and the integrated magnetic current density over the infinitesimal width of the interface region; that integrated quantity is independent of the dielectric function in the interface region. Furthermore, we can express the boundary magnetic current density in terms of the boundary value of the field and obtain a boundary scattering problem. Reviving the retardation effect, the boundary scattering problem is equivalent to the boundary value problem based on (38a)–(38e).

4. To obtain an expression for the boundary magnetic current density we may start from *any smoothing function* that satisfies (61a)–(61b). For example, we may take the next ξ

$$\xi(X) = \frac{1+2X}{2}\epsilon_0 + \frac{1-2X}{2}\epsilon_1 \, , \tag{83}$$

where $X \in \{X|X \in \boldsymbol{R}, -1/2 < X < +1/2\}$. Note that the domain is defined as an open section; outside the section, $\xi(X)$ should be an adequate function of X, which satisfies the conditions (61a)–(61b). Equation (82) together with (83) leads to,

$$\frac{\mathrm{d}}{\mathrm{d}x}D_i(x) = -\frac{\epsilon_1 - \epsilon_0}{\frac{1+2X^\bullet}{2}\epsilon_0 + \frac{1-2X^\bullet}{2}\epsilon_1}\left(\frac{1+2X^\bullet}{2}D_{i0} + \frac{1-2X^\bullet}{2}D_{i1}\right)\delta(x)$$
$$i = 1,2 \qquad x \in \text{ the whole space of interest} \, , \tag{84}$$

where $X^\bullet \in \{X|X \in \boldsymbol{R}, -1/2 < X < +1/2\}$ is the representative position in the interface region. Furthermore, the coefficient of the delta function in the r.h.s. of (84) is a function of $X^\bullet$ and can be extended to a function defined in a certain complex domain of $X^\bullet$ by means of "analytical continuation" [21]. In other words, although the boundary magnetic current density is derived from $\xi(X)$, which is defined in the real and bounded domain of $X^\bullet$, it is effective beyond the initially assumed domain of $X^\bullet$.
If we compare (84) with (41a)–(41f) under the absence of the retardation effect, then we recognize that $\dfrac{1-2X^\bullet}{2}$ corresponds to the arbitrary complex-valued function $\alpha(\boldsymbol{s})$ and the r.h.s. of (84) corresponds to the term of the boundary magnetic current density $-\hat{V}_s[\boldsymbol{C}]$ in (41a)–(41f). Therefore, (84) justifies (41a)–(41f), reviving the retardation effect.
5. Furthermore, one may start from a different smoothing function ξ, the form of which is defined by means of some parameters. Then, one will trace the above procedure and derive a different expression for the boundary magnetic current density, which includes $X^\bullet$ originating from the representative coordinate and the additional parameters introduced in the definition of ξ. Therefore, how to express the boundary magnetic current densityit is rather arbitrary.
In short, the boundary magnetic current density is expressed with a certain arbitrariness originated from the arbitrariness of the profile of the source (or the dielectric function) in the interface region.

The proof is omitted because its procedure is almost the same as that in Sect. 8.1.

9 Green's Function and Delta Function in Vector Field Analysis

Concerning Green's function in vector field analysis, we can refer to [22] or other books. However, there are many notations and expressions; some of them possibly contain errors. Therefore, here we give a concise and self-contained derivation of Green's function. Green's function is useful to solve inhomogeneous Helmholtz equations in the scalar and/or vector version; those equations include our master equations (20a)–(20c) (in the limit of $k \to 0$) and (41a)–(41f). We mention only the Helmholtz equations in the scalar version (85) and the vector version (86)

$$(-\triangle - k^2)X(\boldsymbol{r}) = -V(\boldsymbol{r})\,, \tag{85}$$

$$(-\triangle - k^2)\boldsymbol{X}(\boldsymbol{r}) = -\boldsymbol{V}(\boldsymbol{r})\,. \tag{86}$$

The corresponding Laplace equation can be regarded as the limit of $k \to 0$.

9.1 Vector Helmholtz Equation

Let us start from solving (85); this solution is useful to obtain a solution of (86). Green's function for (85) is defined by (87) and its solution is (88),

$$(-\triangle - k^2)G(\boldsymbol{r}-\boldsymbol{r}';k) = -\delta^3(\boldsymbol{r}-\boldsymbol{r}')\,, \tag{87}$$

$$G(\boldsymbol{r}-\boldsymbol{r}';k) = \frac{-1}{4\pi}\frac{\exp(\pm \mathrm{i}k|\boldsymbol{r}-\boldsymbol{r}'|)}{|\boldsymbol{r}-\boldsymbol{r}'|}\,. \tag{88}$$

We may add any solution of (87) without the source to (88) in order to adjust the boundary condition at $r \to \infty$ or elsewhere, if it is given. In many cases, we take the "+"-signed exponent in (88) because of the causality. Using (88), (85) leads to the next integral form,

$$X(\boldsymbol{r}) = X^{(0)}(\boldsymbol{r}) + \int \mathrm{d}^3r' G(\boldsymbol{r}-\boldsymbol{r}';k)V(\boldsymbol{r}')\,, \tag{89}$$

where $X^{(0)}(\boldsymbol{r})$ is a solution of (85) without the source. If V is a functional of $X(\boldsymbol{r})$, (89) is an integral equation, e.g., (21).

Let us consider (86). The identity tensor (or delta tensor, delta dyadic) is defined through (90) and its solution is (91),

$$\int \mathrm{d}^3r' \mathcal{D}(\boldsymbol{r}-\boldsymbol{r}')_{ij}\boldsymbol{X}(\boldsymbol{r}')_j = \boldsymbol{X}(\boldsymbol{r})_i\,, \tag{90}$$

$$\mathcal{D}(\boldsymbol{r}-\boldsymbol{r}')_{ij} = \delta_{ij}\delta^3(\boldsymbol{r}-\boldsymbol{r}')\,, \tag{91}$$

where $\boldsymbol{X}(\boldsymbol{r})$ is an arbitrary vector field and we use Einstein's rule for the contraction of tensor indices, e.g., $\mathcal{D}_{ij}\boldsymbol{X}_j = \sum_{j=1}^{3}\mathcal{D}_{ij}\boldsymbol{X}_j$.

Then, Green's tensor (or Green's dyadic) is defined by (92) and its explicit expression is (93), because (92) is merely (87) times δ_{ij}.

$$(-\triangle - k^2)\mathcal{G}(\boldsymbol{r}-\boldsymbol{r}';k)_{ij} = -\mathcal{D}(\boldsymbol{r}-\boldsymbol{r}')_{ij} , \tag{92}$$

$$\mathcal{G}(\boldsymbol{r}-\boldsymbol{r}';k)_{ij} = \delta_{ij} G(\boldsymbol{r}-\boldsymbol{r}';k) . \tag{93}$$

Using (93), (86) leads to the next integral form,

$$\boldsymbol{X}(\boldsymbol{r})_i = \boldsymbol{X}^{(0)}(\boldsymbol{r})_i + \int \mathrm{d}^3 r' \mathcal{G}(\boldsymbol{r}-\boldsymbol{r}';k)_{ij} \boldsymbol{V}(\boldsymbol{r}')_j , \tag{94}$$

where $\boldsymbol{X}^{(0)}(\boldsymbol{r})$ is a solution of (86) without the source. If $\boldsymbol{V}$ is a functional of $\boldsymbol{X}(\boldsymbol{r})$, (94) is an integral equation, e.g., (43).

In the following, we call the delta tensor (Green's tensor) the delta function (Green's function) for simplicity.

9.2 Decomposition into Longitudinal and Transversal Components

It is well known that a vector field $\boldsymbol{Y}(\boldsymbol{r})$ can always be decomposed into the two orthogonal components, namely the longitudinal component $\boldsymbol{Y}^{(l)}$ and the transversal one $\boldsymbol{Y}^{(t)}$,

$$\boldsymbol{Y}(\boldsymbol{r}) = \boldsymbol{Y}^{(l)}(\boldsymbol{r}) + \boldsymbol{Y}^{(t)}(\boldsymbol{r}) , \tag{95a}$$

$$\nabla \times \boldsymbol{Y}^{(l)}(\boldsymbol{r}) = \boldsymbol{O} , \tag{95b}$$

$$\nabla \cdot \boldsymbol{Y}^{(t)}(\boldsymbol{r}) = 0 , \tag{95c}$$

$$\int \mathrm{d}^3 r \boldsymbol{Y}^{(l)}(\boldsymbol{r}) \cdot \boldsymbol{Y}^{(t)}(\boldsymbol{r}) = 0 . \tag{95d}$$

If necessary, one can prove the above fact, tracing the procedure from (99) to (106d) with an adequate replacement.

Applying (95a)–(95c) to $\boldsymbol{X}(\boldsymbol{r})$ and $\boldsymbol{V}(\boldsymbol{r})$ in (86) and using an identity,

$$-\triangle \boldsymbol{Y}(\boldsymbol{r}\cdots) = -\nabla\nabla\cdot \boldsymbol{Y}(\boldsymbol{r}\cdots) + \nabla\times\nabla\times \boldsymbol{Y}(\boldsymbol{r}\cdots) , \tag{96}$$

we obtain the longitudinal and the transversal vector Helmholtz equations,

$$(-\nabla\nabla\cdot -k^2)\boldsymbol{X}^{(l)}(\boldsymbol{r}) \quad = -\boldsymbol{V}^{(l)}(\boldsymbol{r}) , \tag{97}$$

$$(\nabla\times\nabla\times -k^2)\boldsymbol{X}^{(t)}(\boldsymbol{r}) = -\boldsymbol{V}^{(t)}(\boldsymbol{r}) . \tag{98}$$

Each equation is natural because the longitudinal (transversal) source yields the longitudinal (transversal) field.

Note that the decomposition of the equation is not effective in (86) under $k \to 0$, namely the vector Laplace equation. A solution of the vector Laplace equation does not propagate because of $k \to 0$ and the concept of longitudinal or transversal components of a solution is meaningless. In fact, the expressions for $\mathcal{G}^{(l)}$ and $\mathcal{G}^{(t)}$, i.e., (112) and (113), possess a certain singularity in the limit of $k \to 0$, while (88) and (93) do not.

Equations (97) and (98) lead to integral forms, if the corresponding longitudinal and transversal vector Green's functions are defined; we will define these in the following.

In general, a diagonal tensor field is defined as the product of Kronecker's delta and a scalar field that depends only on the difference between the two spatial coordinates,

$$\mathcal{Y}(\boldsymbol{r}-\boldsymbol{r}')_{ij} \equiv \delta_{ij} Y(\boldsymbol{r}-\boldsymbol{r}') \,. \tag{99}$$

A diagonal tensor can always be decomposed into the longitudinal and the transversal tensor components. Note that $\mathcal{D}$ and $\mathcal{G}$ belong to such a class of tensor.

$$\begin{aligned}\mathcal{Y}(\boldsymbol{r}-\boldsymbol{r}')_{ij} &= \delta_{ij} Y(\boldsymbol{r}-\boldsymbol{r}') \\ &= \delta_{ij}\frac{1}{(2\pi)^3}\int \mathrm{d}^3k \tilde{Y}(\boldsymbol{k}) \exp\left(+\mathrm{i}\boldsymbol{k}\cdot(\boldsymbol{r}-\boldsymbol{r}')\right) \\ &= \frac{1}{(2\pi)^3}\int \mathrm{d}^3k \left(\hat{\boldsymbol{k}}_i\hat{\boldsymbol{k}}_j + \hat{\mathbf{e}}_i^{(2)}\hat{\mathbf{e}}_j^{(2)} + \hat{\mathbf{e}}_i^{(3)}\hat{\mathbf{e}}_j^{(3)}\right) \\ &\qquad\qquad \tilde{Y}(\boldsymbol{k}) \exp\left(+\mathrm{i}\boldsymbol{k}\cdot(\boldsymbol{r}-\boldsymbol{r}')\right) \,, \end{aligned} \tag{100}$$

where $\tilde{Y}(\boldsymbol{k})$ is the Fourier transformation of $Y(\boldsymbol{r})$; we implicitly assume that the Fourier integral is converged. To obtain the last expression in (100), we use the expansion of δ_{ij} by a normalized orthogonal complete set $\{\hat{\mathbf{e}}^{(\alpha)}|\alpha = 1,2,3\}$,

$$\delta_{ij} = \hat{\mathbf{e}}_i^{(1)}\hat{\mathbf{e}}_j^{(1)} + \hat{\mathbf{e}}_i^{(2)}\hat{\mathbf{e}}_j^{(2)} + \hat{\mathbf{e}}_i^{(3)}\hat{\mathbf{e}}_j^{(3)} \,. \tag{101}$$

Furthermore, we take one of the bases as the unit vector of the wavenumber vector $\hat{\mathbf{e}}^{(1)} = \hat{\boldsymbol{k}}$ in the integrand. Now let us define the first term (the remaining terms) in the last expression in (100) as $\mathcal{Y}^{(l)}(\boldsymbol{r}-\boldsymbol{r}')_{ij}$ ($\mathcal{Y}^{(t)}(\boldsymbol{r}-\boldsymbol{r}')_{ij}$). We recognize $\mathcal{Y}^{(l)}(\boldsymbol{r}-\boldsymbol{r}')$ as the longitudinal tensor because its integrand is composed of the direct product of the wavenumber vector $\boldsymbol{k}$'s, while $\mathcal{Y}^{(t)}(\boldsymbol{r}-\boldsymbol{r}')_{ij}$ is the transversal tensor because $\hat{\mathbf{e}}^{(2)}$ and $\hat{\mathbf{e}}^{(3)}$ in the integrand are orthogonal to the wavenumber vector $\boldsymbol{k}$.

As a result, any diagonal tensor field $\mathcal{Y}$ defined by (99) (including the cases of $\mathcal{D}$ and $\mathcal{G}$) is decomposed into the longitudinal and the transversal components,

$$\mathcal{Y}(\boldsymbol{r}-\boldsymbol{r}')_{ij} = \mathcal{Y}^{(l)}(\boldsymbol{r}-\boldsymbol{r}')_{ij} + \mathcal{Y}^{(t)}(\boldsymbol{r}-\boldsymbol{r}')_{ij} \,, \tag{102a}$$

$$\nabla \times \mathcal{Y}^{(l)}(\boldsymbol{r}-\boldsymbol{r}') = \nabla' \times \mathcal{Y}^{(l)}(\boldsymbol{r}-\boldsymbol{r}') = \mathcal{O} \,, \tag{102b}$$

$$\nabla \cdot \mathcal{Y}^{(t)}(\boldsymbol{r}-\boldsymbol{r}') = \nabla' \cdot \mathcal{Y}^{(t)}(\boldsymbol{r}-\boldsymbol{r}') = \mathbf{0} \,. \tag{102c}$$

The detailed expression for (102b) using the antisymmetric tensor ϵ_{ijk} is

$$\epsilon_{ijk}\partial_j \mathcal{Y}^{(l)}(\boldsymbol{r}-\boldsymbol{r}')_{kl} = \epsilon_{ijk}\partial'_j \mathcal{Y}^{(l)}(\boldsymbol{r}-\boldsymbol{r}')_{lk} = O \,. \tag{103}$$

Now the vector delta function (91) and vector Green's function (93) are the diagonal tensor and can be decomposed as $\mathcal{D} = \mathcal{D}^{(l)} + \mathcal{D}^{(t)}$, and $\mathcal{G} = \mathcal{G}^{(l)} + \mathcal{G}^{(t)}$, respectively.

Let us deduce explicit expressions for $\mathcal{D}^{(l)}$, $\mathcal{D}^{(t)}$, $\mathcal{G}^{(l)}$, and $\mathcal{G}^{(t)}$. The Fourier representation of the vector delta function (91) is (100) with $\tilde{Y}(\boldsymbol{k}) = 1$. Then, $\mathcal{D}^{(l)}$ and $\mathcal{D}^{(t)}$ is calculated as follows,

$$\begin{aligned}\mathcal{D}^{(l)}(\boldsymbol{r}-\boldsymbol{r}')_{ij} &\equiv \frac{1}{(2\pi)^3}\int \mathrm{d}^3k\hat{\boldsymbol{k}}_i\hat{\boldsymbol{k}}_j \exp\left(+\mathrm{i}\boldsymbol{k}\cdot(\boldsymbol{r}-\boldsymbol{r}')\right)\\ &= \partial_i\partial'_j\frac{1}{(2\pi)^3}\int \mathrm{d}^3k\frac{1}{k^2}\exp\left(+\mathrm{i}\boldsymbol{k}\cdot(\boldsymbol{r}-\boldsymbol{r}')\right)\\ &= \partial_i\partial'_j\frac{1}{2\pi^2}\int_0^\infty \mathrm{d}k\frac{\sin(k|\boldsymbol{r}-\boldsymbol{r}'|)}{k|\boldsymbol{r}-\boldsymbol{r}'|}\\ &= -\partial_i\partial'_j G(\boldsymbol{r}-\boldsymbol{r}';0)\,. \end{aligned} \tag{104}$$

The 3rd part of (104) is deduced using the fact that "i$\boldsymbol{k}$" is equivalent to the operator ∇ or $-\nabla'$ in the integrand and the 4th part is the result of the integration over the angular variables in the spherical coordinates. The last expression is derived by the use of $\int_0^\infty \sin x/x = \pi/2$. On the other hand,

$$\begin{aligned}\mathcal{D}^{(t)}(\boldsymbol{r}-\boldsymbol{r}')_{ij} &\equiv \mathcal{D}(\boldsymbol{r}-\boldsymbol{r}')_{ij} - \mathcal{D}^{(l)}(\boldsymbol{r}-\boldsymbol{r}')_{ij}\\ &= \delta_{ij}\delta^3(\boldsymbol{r}-\boldsymbol{r}') + \partial_i\partial'_j G(\boldsymbol{r}-\boldsymbol{r}';0)\,. \end{aligned} \tag{105}$$

We can check that $\mathcal{D}^{(l)}$ and $\mathcal{D}^{(t)}$ is the identity or the projection operators in the subspace of the longitudinal and the transversal vector field, respectively,

$$\int \mathrm{d}^3r'\mathcal{D}^{(l)}(\boldsymbol{r}-\boldsymbol{r}')_{ij}\boldsymbol{Y}^{(l)}(\boldsymbol{r}')_j = \boldsymbol{Y}^{(l)}(\boldsymbol{r})_i\,, \tag{106a}$$

$$\int \mathrm{d}^3r'\mathcal{D}^{(l)}(\boldsymbol{r}-\boldsymbol{r}')_{ij}\boldsymbol{Y}^{(t)}(\boldsymbol{r}')_j = \boldsymbol{0}\,, \tag{106b}$$

$$\int \mathrm{d}^3r'\mathcal{D}^{(t)}(\boldsymbol{r}-\boldsymbol{r}')_{ij}\boldsymbol{Y}^{(t)}(\boldsymbol{r}')_j = \boldsymbol{Y}^{(t)}(\boldsymbol{r})_i\,, \tag{106c}$$

$$\int \mathrm{d}^3r'\mathcal{D}^{(t)}(\boldsymbol{r}-\boldsymbol{r}')_{ij}\boldsymbol{Y}^{(l)}(\boldsymbol{r}')_j = \boldsymbol{0}\,. \tag{106d}$$

Now we can calculate $\mathcal{G}^{(l)}$ and $\mathcal{G}^{(t)}$. Applying (102a)–(102c) to $\mathcal{D}$ and $\mathcal{G}$ in (86) and using (96), (86) is decomposed into the next two equations for the longitudinal and the transversal components,

$$(-\nabla\nabla\cdot -k^2)\mathcal{G}^{(l)}(\boldsymbol{r}-\boldsymbol{r}';k) = -\mathcal{D}^{(l)}(\boldsymbol{r}-\boldsymbol{r}')\,, \tag{107}$$

$$(\nabla\times\nabla\times -k^2)\mathcal{G}^{(t)}(\boldsymbol{r}-\boldsymbol{r}';k) = -\mathcal{D}^{(t)}(\boldsymbol{r}-\boldsymbol{r}')\,. \tag{108}$$

From (107), we obtain an explicit expression for $\mathcal{G}^{(l)}$ as follows,

$$\begin{aligned}
\mathcal{G}^{(l)}(\boldsymbol{r}-\boldsymbol{r}';k)_{ij} &= \frac{1}{k^2}\left(-\nabla\nabla\cdot\mathcal{G}^{(l)}(\boldsymbol{r}-\boldsymbol{r}';k)+\mathcal{D}^{(l)}(\boldsymbol{r}-\boldsymbol{r}')\right)_{ij} \\
&= \frac{1}{k^2}\left(-\nabla\nabla\cdot\mathcal{G}(\boldsymbol{r}-\boldsymbol{r}';k)+\mathcal{D}^{(l)}(\boldsymbol{r}-\boldsymbol{r}')\right)_{ij} \\
&= \frac{1}{k^2}\left(-\partial_i\partial_k\delta_{kj}G(\boldsymbol{r}-\boldsymbol{r}';k)-\partial_i\partial_j'G(\boldsymbol{r}-\boldsymbol{r}';0)\right) \\
&= \frac{1}{k^2}\partial_i\partial_j'\left(G(\boldsymbol{r}-\boldsymbol{r}';k)-G(\boldsymbol{r}-\boldsymbol{r}';0)\right) . \qquad (109)
\end{aligned}$$

Furthermore, an expression for the transversal vector Green's function is

$$\begin{aligned}
\mathcal{G}^{(t)}(\boldsymbol{r}-\boldsymbol{r}';k)_{ij} &= \mathcal{G}(\boldsymbol{r}-\boldsymbol{r}';k)_{ij}-\mathcal{G}^{(l)}(\boldsymbol{r}-\boldsymbol{r}';k)_{ij} \qquad (110)\\
&= \delta_{ij}G(\boldsymbol{r}-\boldsymbol{r}';k)-\frac{1}{k^2}\partial_i\partial_j'\left(G(\boldsymbol{r}-\boldsymbol{r}';k)-G(\boldsymbol{r}-\boldsymbol{r}';0)\right) .
\end{aligned}$$

As a result, we can convert (97) and (98) to the next two integral forms using (106a)–(106d), (107) and (108),

$$\boldsymbol{X}^{(l/t)}(\boldsymbol{r})_i = \boldsymbol{X}^{(l/t)(0)}(\boldsymbol{r})_i + \int \mathrm{d}^3r'\mathcal{G}^{(l/t)}(\boldsymbol{r}-\boldsymbol{r}';k)_{ij}\boldsymbol{V}^{(l/t)}(\boldsymbol{r}')_j . \qquad (111)$$

For a numerical calculation, the next expressions are convenient,

$$\begin{aligned}
\mathcal{G}^{(l)}(\boldsymbol{R};k)_{ij} &= -\frac{G(\boldsymbol{R};0)}{(ikR)^2}(-\delta_{ij}+3\hat{\boldsymbol{R}}_i\hat{\boldsymbol{R}}_j) \qquad (112)\\
&\quad +\frac{G(\boldsymbol{R};k)}{(ikR)^2}\left(-\delta_{ij}(1-ikR)+\hat{\boldsymbol{R}}_i\hat{\boldsymbol{R}}_j(3-3ikR+(ikR)^2)\right) , \\
\mathcal{G}^{(t)}(\boldsymbol{R};k)_{ij} &= +\frac{G(\boldsymbol{R};0)}{(ikR)^2}(-\delta_{ij}+3\hat{\boldsymbol{R}}_i\hat{\boldsymbol{R}}_j) \qquad (113)\\
&\quad -\frac{G(\boldsymbol{R};k)}{(ikR)^2}\left(-\delta_{ij}(1-ikR+(ikR)^2)\right. \\
&\quad \left.+\hat{\boldsymbol{R}}_i\hat{\boldsymbol{R}}_j(3-3ikR+(ikR)^2)\right) ,
\end{aligned}$$

where $\boldsymbol{R}\equiv\boldsymbol{r}-\boldsymbol{r}'$ and $\hat{\boldsymbol{R}}=\boldsymbol{R}/|\boldsymbol{R}|$. The expressions in [22] corresponding to (112) and (113) are not correct.

References

1. E.H. Synge: Philos. Mag. **6**, 356 (1928)
2. J.A. O'Keefe: J. Opt. Soc. Am. **46**, 359 (1956)
3. E.A. Ash, G. Nicholls: Nature **237**, 510 (1972)
4. G.A. Massey, J.A. Davis, S.M. Katnik, E. Omon: Appl. Opt. **24**, 1498 (1985)
5. D.W. Pohl, W. Denk, M. Lanz: Appl. Phys. Lett. **44**, 651 (1984)

6. E. Betzig, M. Issacson, A. Lewis: Appl. Phys. Lett. **51**, 2088 (1987)
7. S. Jiang, H. Ohsawa, K. Yamada, T. Pangaribuan, M. Ohtsu, K. Imai, A. Ikai: Jpn. J. Appl. Phys. **31**, 2282 (1992)
8. *Near-field Nano/Atom Optics and Technology* ed. by M. Ohtsu (Springer, Tokyo 1998)
9. H.A. Bethe: Phys. Rev. **66**, 163 (1944)
10. C.J. Bouwkamp: Philips Res. Rep. **5**, 401 (1950)
11. C. Girared, D. Coujon: Phys. Rev. B **42**, 9340 (1990)
12. J.F. Martin, C. Girard, A. Dereux: Phys. Rev. Lett. **74**, 526 (1995)
13. K.S. Yee: IEEE Trans. Antennas Prop. **AP14**, 302 (1966)
14. I. Banno, H. Hori, T. Inoue: Opt. Rev. **3**, 454 (1996)
15. I. Banno, H. Hori: 'Dual Ampere Law and Image of NOM' (in Japanese). In: *Handbook for Near-field Nano-photonics* ed. by M. Ohtsu, S. Kawata (Optronics, Tokyo 1996) pp. 240–244, ibid. In: *Introduction to Near-field Nano-photonics* ed. by M. Ohtsu, S. Kawata (Optronics, Tokyo 2000) pp. 48–52
16. I. Banno, H. Hori: Trans. IEE Japan **119-C**, 1094 (1999) (in Japanese)
17. R.P. Feynmann, R.B. Leighton, M.L. Sands: *The Feynman Lectures on Physics, vol.2* (Addison-Wesley, 1965)
18. J.A. Stratton: *Electromagnetic Theory* (McGraw-Hill, New York, London 1941)
19. M. Naya, S. Mononobe, R. Uma Maheswari, T. Saiki, M. Ohtsu: Opt. Commun. **124**, 9 (1996)
20. W. Jhe, K. Jang: Ultramicroscopy **61**, 81 (1995)
21. T. Takagi: *Kaiseki Gairon (Introduction to Analysis) 3rd edn* (Iwanami-Shoten, Tokyo 1983) (in Japanese)
22. P.M. Morse, H. Feshbach: *Methods of Theoretical Physics* (McGraw-Hill, New York, Toronto, London 1953)

Excitonic Polaritons in Quantum-Confined Systems and Their Applications to Optoelectronic Devices

T. Katsuyama and K. Hosomi

1 Introduction

The dynamic behavior of excitons in semiconductors has become an attractive field of research since it was found to play an important role in the optical transitions near the fundamental absorption edge of semiconductors [1–3]. One of the most interesting phenomena related to excitons is represented by the excitonic polaritons, which are coherent quasiparticles consisting of photons and excitons. The concept of the excitonic polariton was introduced by Pekar [4] and Hopfield [5,6]. Such materials as CuCl [7], CdS [8], and GaAs [9] have been used to experimentally study the actual phenomena. The existence of excitonic polaritons has been clearly shown at liquid-helium temperature by time-of-flight measurements obtained with laser pulses propagating in semiconductors [10,11] and also by resonant Brillouin scattering [8,9] and two-photon resonant Raman scattering [7,12]. It has also become possible to fabricate systems that consist of two different semiconductors in alternating layers of controlled thickness with relatively sharp interfaces. These systems can be grown by using such epitaxial crystal-growth techniques as molecular beam epitaxy (MBE) and metalorganic chemical vapor deposition (MOCVD). Much effort has recently been devoted to the study of carrier dynamics in these artificial layered structures such as quantum wells and superlattices. Since the band structure depends critically on the composition, it is modulated in the direction normal to the layers. On the other hand, along the plane of the layers, the band structure is not very different from that of the parent compounds. This leads to the highly anisotropic electronic and optical properties that are inherent to quantum wells and superlattices.

Exciton dynamics in these quantum wells and superlattices is also important in understanding such anisotropic properties. In particular, the increase in the exciton binding energy due to the quantum-confinement effect [13] has opened up the possibility of optical device applications because exceptionally clear exciton resonance absorption even at room temperature can be seen in GaAs/AlGaAs quantum wells [14]. Therefore, the fast phenomena that are related to excitonic processes, i.e., the quantum-confined Stark effect [15,16], the ac Stark effect [17,18], and many of the nonlinear effects that arise under laser-pulse irradiation [19], have been subject to extensive study.

Excitonic polaritons in such quantum-confined systems are an interesting subject for research. The luminescence and reflection spectra of low-temperature GaAs/AlGaAs quantum-well structures have been studied and the existence of the excitonic polariton has been discussed [20,21]. Prior to these experiments, a theoretical investigation of quantum-well excitonic polaritons predicted the enhancement of photon–exciton interaction due to the quantum-confinement effect, resulting in the increase of the stability of the excitonic polariton [22].

Furthermore, the propagation of quantum-well excitonic polaritons in a waveguide structure has been demonstrated by using time-of-flight measurements [23,24]. The propagation of quantum-well excitonic polaritons in a waveguide structure has also been proven by the first-principles calculation of the spatial-dispersion relations of excitonic polaritons [25]. This calculation was based on an additional-boundary-condition-free theory [26] and guided-mode characteristics. The waveguide structure offers artificial control of the coupling between the exciton and photon fields by changing the geometrical configuration of the waveguide [27,28]. A further interesting aspect of the excitonic polariton is related to the cavity effect [29]. Introducing a grating along the waveguide is a way of improving the coupling between the exciton and photon fields, i.e., the polariton stability. Thus, the characteristics of the excitonic polariton characteristics itself can be subjected to artificial control [30].

Furthermore, the waveguide structure makes it possible to guide the excitonic polaritons in a way that is similar to that in a conventional guided-wave optical device [30]. For example, a monolithic interferometer using a waveguide structure can be fabricated, thus achieving an excitonic-polariton-based interferometer. In addition, the electric-field-induced phase change of the excitonic polariton in the waveguide is significantly enhanced since the electric field distorts the exciton wavefunction and effectively modifies the photon–exciton interaction [31–33]. Therefore, the output light from the polariton interferometer can be sensitively modulated by applying an electric field [34]. Such a device configuration inevitably offers advanced optical devices such as optical switches with an extremely low-voltage operation, which leads to the low power dissipation and high operation drivability: these are quite useful for applications to the present optical communications networks with high bit rates [35–38].

Another advantage of excitonic polariton propagation is that the modal field in the waveguide may be spatially squeezed because the effective wavelength in the waveguide is reduced by the increase in the refractive index that arises because of the polariton effect [39]. This advantage offers an ultra-thin waveguide, from which it is possible to construct various waveguide-type optical devices with extremely small dimensions that reach down into the nanometer scale [40]. This is expected to lead to new devices that will allow the realization of such advanced optical systems as those for optical interconnection in LSI chips. Thus, devices in which polariton propagation is applied have the potential to open up a new device paradigm in fields ranging from the systems for the optical communication network of the future to extremely small systems of optical interconnection [30,36].

In this chapter, we review the fundamental aspects of excitonic polaritons that are propagating in quantum-well waveguides. We then discuss the possible applications of these excitonic polaritons to optoelectronic devices. The main part of this article is organized in two sections. Section 2 presents fundamental aspects of excitonic polaritons propagating in quantum-well waveguides. In this section, experimental results [23,24,27] and theoretical calculations [25] are given, which verify the existence of excitonic polaritons in quantum-well waveguides. In addition, the polariton behaviors under an electric field, particularly phase characteristics of the excitonic polariton, are presented on the basis of the experimental results [31–34,38]. The cavity effects of excitonic polaritons are also discussed in terms of polariton stability [29].

Section 3 describes the possible applications of propagation of these excitonic polaritons in such quantum-confined systems as quantum wells and wires. The applications are particularly oriented to optoelectronic devices. First, two devices, i.e., Mach—Zehnder-type modulators and directional-coupler-type switches are described [37,38]. These devices are expected to reduce the voltages for the operation, which lead to lower levels of power dissipation and high drivability in operation. Approaches to nanometer-scale optical devices are then presented. The confinement of light fields in extremely small waveguides is discussed [39]. Reduction of the optical switch size to the nanometer scale is also discussed [40]. Finally, a summary is given and some future prospects for research into the fundamental aspects and applications of excitonic polaritons are presented.

2 Fundamental Aspects of Excitonic Polaritons Propagating in Quantum-Confined Systems

2.1 The Concept of the Excitonic Polariton

Prior to our discussion of excitonic polaritons in quantum wells, we will briefly review the concept of the excitonic polariton [2]. The exciton has a polarization with longitudinal and transverse modes. Only the transverse exciton can interact with an electromagnetic field. This dipole interaction splits the exciton into longitudinal and transverse branches and produces nonanalytic behavior as $k \to 0$ (since the distinction between longitudinal and transverse ceases to exist at $k = 0$). The effect of this interaction is shown in Fig. 1, which is a schematic energy and wavevector diagram for the longitudinal and transverse excitons and for the photon (dashed line). The photon and the transverse exciton become mixed in the crossover region, losing their identity

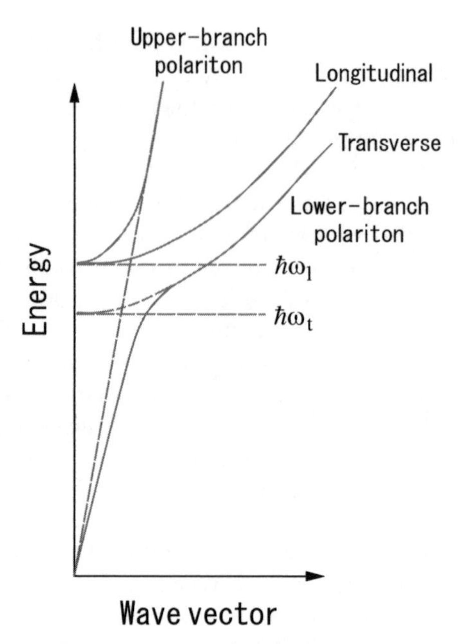

Fig. 1. Schematic view of the dispersion (energy E, wave vector k) relation of the excitonic polariton. $\hbar\omega_l$ and $\hbar\omega_t$ are the longitudinal exciton energy at $k = 0$ and transverse exciton energy at $k = 0$, respectively

in a combined particle called an "excitonic polariton", as shown in Fig. 1. The most important feature of this excitonic polariton is that the lower branch curves upwards and two different polariton states (upper-branch polariton and lower-branch polariton) can coexist at levels above the longitudinal exciton energy defined at $k = 0$. Such a dispersion relation leads to a k-dependent nonlocal dielectric function, a phenomenon known as "spatial dispersion".

The concept of the excitonic-polariton may be represented by the difference between the longitudinal exciton energy $\hbar\omega_l$ and the transverse exciton energy $\hbar\omega_t$. This difference is called the longitudinal-transverse (LT) splitting energy Δ. If the LT splitting energy is much lower than the thermal energy, the excitonic-polariton effect vanishes and only the optical effect remains. This means that the dispersion relation of the excitonic polariton changes to the conventional dispersion relation of the light by the decrease in LT splitting.

Time-of-flight measurement techniques provide one of the most reliable methods for proving the existence of the excitonic polariton [10,11]. Since the curvature of the polariton dispersion around the exciton resonance is quite steep, the group velocity is significantly changed, resulting in a large delay in the transmitted light pulse. In the following section, this method is used to prove that the excitonic polariton can propagate in a quantum-well waveguide.

2.2 Excitonic Polaritons in GaAs Quantum-Well Waveguides: Experimental Observations

This section describes time-of-flight measurements used to show the existence of excitonic polaritons propagating in a GaAs quantum-well waveguide [23,24].

Since the excitonic polariton is a complex particle that consists of a photon and an exciton, as was explained in the previous section, it must propagate in the direction parallel to the quantum well [22]. This is because a translational motion of the quantum-well excitons along the layers is possible. Therefore, a waveguide-type sample that contains a single GaAs quantum well should provide a way of proving the existence of the quantum-well excitonic polariton.

The sample used in the experiments [23,24] was grown by molecular beam epitaxy on a semi-insulating GaAs substrate. The sample structure is shown in Fig. 2. The sample forms a leaky waveguide, in which the refractive index of the core is smaller than that of the cladding [41]. A single 50-Å GaAs quantum well is sandwiched between 1.8-μm superlattice layers. Each superlattice layers consists of 250 periods of 30-Å GaAs and 40-Å $Al_{0.3}Ga_{0.7}As$. These layers form the core of the waveguide sample. The cladding is a 1.0-μm GaAs layer. The waveguide is designed to only allow transmission of the fundamental mode. The length of the waveguide is 250 μm.

In order to show the existence of excitonic polariton propagation, a picosecond time-of-flight method is used to measure the propagation delay time of the light transmitted along the quantum-well layer. The experimental arrangement for low-temperature time-of-flight measurement of the quantum-well waveguide is shown in Fig. 3.

Figure 4 shows the dependence, at 6 K, of the excitonic-absorption spectrum on polarization in the quantum-well waveguide. Here, in the transverse-electric (TE) polarization, the electric field of the laser pulse is parallel to the quantum-well layer, whereas for the transverse-magnetic (TM) polarization, the electric field is perpendicular to the quantum-well layer. The laser pulses

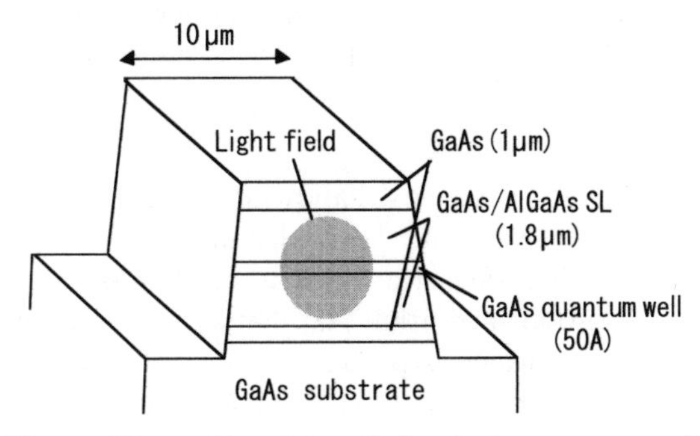

Fig. 2. Waveguide with a GaAs single quantum well

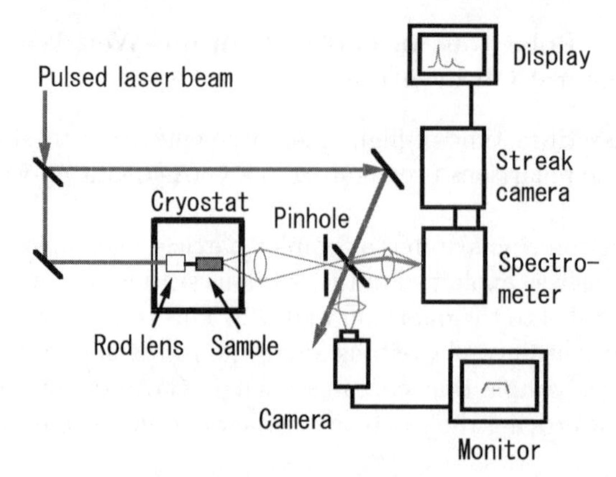

Fig. 3. Experimental arrangement for the low-temperature time-of-flight measurement

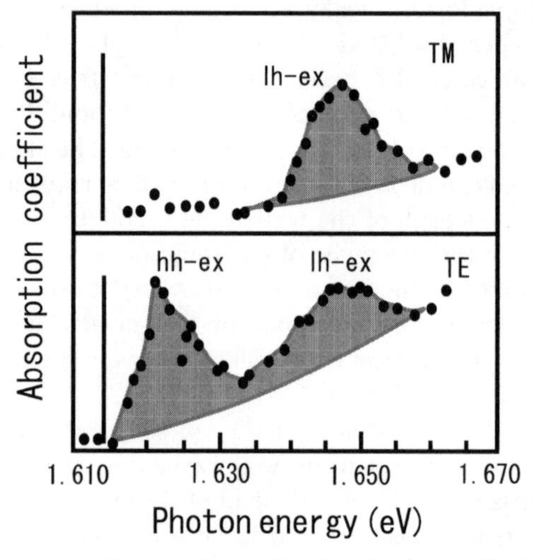

Fig. 4. Absorption spectra for two-dimensional excitons in a GaAs single-quantum-well waveguide in TE (*lower*) and TM (*upper*) polarization at 6 K. *Shaded areas* of each spectrum correspond to exciton absorption. The labels hh-ex and lh-ex indicate the heavy-hole and light-hole exciton, respectively. Each *vertical bar* indicates the absorption coefficient $\alpha_0 = 190\ \mathrm{cm}^{-1}$

are 5 ps long and have a 0.85-meV bandwidth. The power density of the incident light is $8.5 \times 10^{-10}\ \mathrm{J/cm^2}$ for each pulse (3.3×10^{10} photons/cm^2 pulse), which is more than one order of magnitude less than that required for the onset of absorption saturation [42]. By assuming an exponential tail for the background part, which is probably due to the band-tail absorption by the superlattice layers, the excitonic-absorption lines are obtained as the shaded area for each polarization. In TE polarization, two absorption lines are observed.

Of these two lines, the one that lies at the lower photon energy, disappears in the TM polarization, while the line at the higher photon energy grows larger in the TM polarization. This feature is qualitatively consistent with the optical anisotropy of heavy-hole (hh) and light-hole (lh) excitons [41]. The line at the lower photon energy corresponds to a hh exciton. The line at the higher photon energy is due to an lh exciton. The anisotropy of the lh exciton is weaker than expected. This weakness may be due to the valence-band mixing in the exciton wavefunction [41].

The absorption coefficient of the waveguide at the photon energy of the hh exciton absorption is found to be $\alpha_0 = 190\ \mathrm{cm}^{-1}$ in the linear regime. The coefficient α_0 is indicated by the vertical bar in each of the parts of Fig. 4. Subtracting the background, the coefficient $\alpha_{\mathrm{hh-ex}}$ for the hh exciton is $170\ \mathrm{cm}^{-1}$. The value of $\alpha_{\mathrm{hh-ex}}$ is about 85% of the value estimated for the room-temperature exciton absorption in GaAs quantum wells [41]. This decrease in the coefficient reflects the coherent propagation of the light pulse.

The time-of-flight spectra for the two polarizations in the exciton-absorption region, at 6 K, are plotted in Fig. 5. The delay time in the TE polarization reaches its maximum at a photon energy that is almost resonant with the hh exciton absorption. The delay time is measured relative to the time of flight of a probe pulse at a photon energy that is far below the exciton absorption. The dispersion curves of excitonic polaritons change drastically from photon-like to exciton-like and vice versa in the energy region of the exciton absorption.

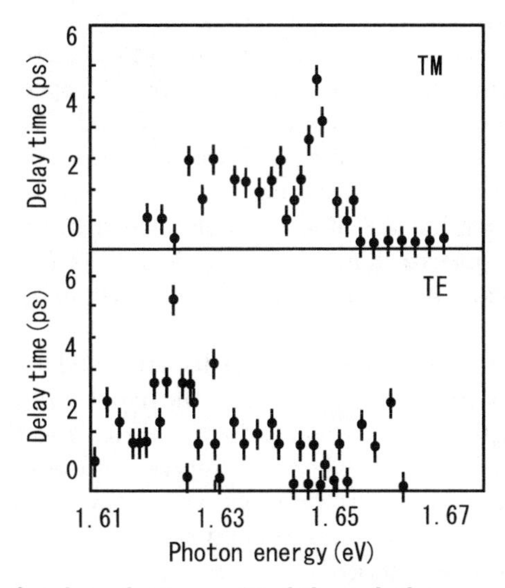

Fig. 5. Delay time of probe pulse transmitted through the waveguide sample at 6 K. The *lower part* corresponds to the delay-time spectra obtained in TE polarization and the *upper part* to the spectra for TM polarization

In general, the group velocity of the polaritons is obtained from the curvature of excitonic polariton dispersion relations. Therefore, this gives a considerable change in the polariton group velocity. The observed peak in the delay time is caused by a decrease in the group velocity of the quantum-well excitonic polaritons, which is associated with the hh excitons.

To derive the group velocity of quantum-well excitonic polaritons in the waveguide, it should be noted that the quantum-well excitonic polaritons propagate as guided waves. The strength of the polariton coupling is determined by a filling factor [41], which corresponds to the overlap between the exciton wave function and the light field. The optical constants of quantum-well excitonic polaritons, such as their absorption coefficient and group velocity, must be divided by this filling factor for comparison with those of bulk excitonic polaritons. From the standard waveguide theory [43], it can be shown that the mode profile of the light field is close to Gaussian, so the filling factor is estimated to be 7.8×10^{-4}. The group velocity of the polariton associated with the hh exciton has been observed to be 3.7×10^{4} m/s, which is almost 1/2300 of the velocity of light in GaAs. The group velocity of polaritons in bulk GaAs at 1.3 K has been reported to be about 1/600 of the velocity of light [11]. The observed decrease in the group velocity of the quantum-well excitonic polaritons is due to the influence of confinement on the oscillator strength compared to the case of excitonic polaritons in the bulk material.

The time delay due to the lh-exciton-associated polariton is almost zero for the light in the TE polarization. This is probably because the TE component of the lh-exciton oscillator strength is almost a quarter of the hh exciton. In the TM polarization, however, a significant increase in the delay time is observed at a photon energy that is resonant with the lh-exciton absorption. The maximum delay time is larger than would be expected from the ratio between lh-exciton oscillator strengths in the TE and TM polarizations. The polariton group velocity associated with the TM lh exciton is 4.2×10^{4} m/s. The light fields extend beyond the narrow quantum-well layer, whereas the exciton wave function has an almost δ-functional distribution in the waveguide space in TM polarization, so the physical situations for the excitonic polariton propagation in TM polarization are very different from that in TE polarization.

Theoretical estimates suggest that the dispersion relation of quantum-well excitonic polaritons is strongly modified by the depolarization fields produced by the excitons in the TM configuration [22]. From the theory [22], we expect a shift of $12(a_{\mathrm{ex}}/L_{\mathrm{W}})\Delta$, where a_{ex} represents the Bohr radius of bulk excitons, L_{W} is the width of the quantum well, and Δ is the longitudinal-transverse splitting of the bulk excitonic polariton, in the depolarization energy for these excitonic polaritons. In the present sample, $L_{\mathrm{W}} = 50$ Å, and the depolarization shift as estimated by using the values $a_{\mathrm{ex}} = 136$ Å and $\Delta = 0.086$ meV [44] is more than 2 meV. This shift is not apparent in the absorption spectrum of Fig. 4. This may be because of mixing between TE and TM polarization. The depolarization effect also shifts the bottom of the parabolic dispersion of the lh exciton from the Γ point to some point with $k \neq 0$, and the exciton group velocity becomes slower. This effect is much stronger on the group velocity than on the energy shift, and qualitatively explains the observed enhancement of the group velocity decrease in TM polarization.

It may be argued that the decrease in the group velocity is due to a change in the refractive index. However, the anomalous dispersion that appears in the spectral region of the exciton resonance would only be capable of causing an increase or at most a negligible decrease in the group velocity. Thus, the refractive index is not considered to have a significant effect on the observed pulse delay.

When the incident power density is increased to 2.1×10^{-8} J/cm^2 per pulse (8.1×10^{12} photons/cm^2 pulse), a dramatic change in the pulse delay time is seen in the spectral region of the exciton absorption. This power density is just above the onset of the absorption saturation [42]. Figure 6 shows an example of the dependence on power of the absorption spectra and the time of flight for the polariton associated with the hh exciton in TE polariza-

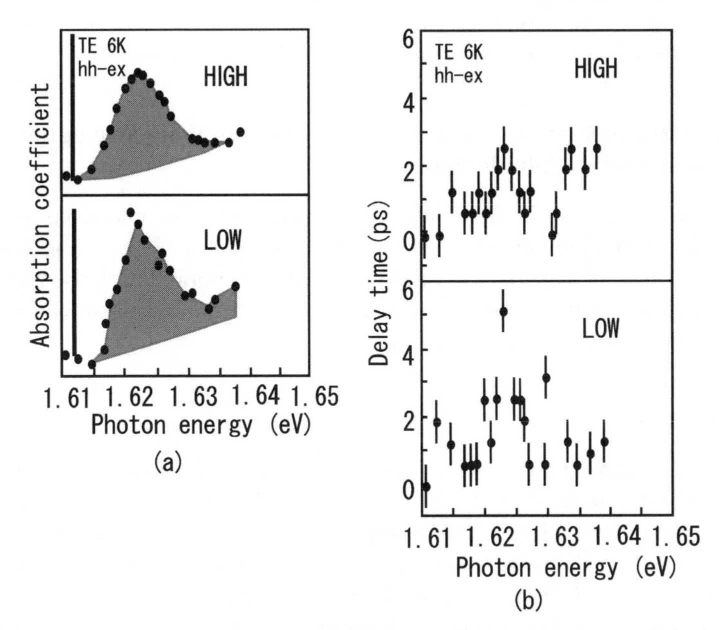

Fig. 6a,b. (**a**) Power dependences of the absorption spectrum and (**b**) the delay time due to quantum-well excitonic polariton propagation for the hh exciton in TE polarization at 6 K. Incident power density for the index 'LOW' is 8.5×10^{-10} J/cm^2 pulse, while that for the index 'HIGH' is 2.1×10^{-8} J/cm^2 pulse. Vertical bars indicate α_0, as in Fig. 4

tion. The intensity of the hh-exciton-absorption line decreases when the laser power increases. The amount of this saturation is $\sim 20\%$ of the intensity of the line measured in the linear regime. However, the pulse-propagation delay times decrease to about half their values in the linear regime; thus, the group velocity has been increased by a factor of 2. This is the opposite tendency to that seen in absorption saturation in relation to the hole-burning effect [42]. According to the relation $v_{\mathrm{g}} \propto 1/(-\Delta\alpha_{\mathrm{ex}})$, which applies to absorption saturation [42], the group velocity v_{g} of the light should decrease with increasing saturation $-\Delta\alpha_{\mathrm{ex}}$. The observed increase in the group velocity is caused by nonlinearity of the quantum-well excitonic polaritons rather than by saturation. The higher incident power generates a larger polariton population. The increased amount of interaction among excitonic polaritons in a larger population leads to increased scattering of the polaritons. The scattering weakens the coupling between the exciton and the photon, so that the polariton dispersion becomes more photon-like. Similar nonlinearity is observed for polaritons associated with the lh exciton.

In summary, the polariton group velocity associated with the heavy-hole exciton was found to be 3.7×10^4 m/s, whereas the velocity associated with the light-hole exciton is 4.2×10^4 m/s. The polariton effect is significantly enhanced in the configuration where the polarization of the incident light is perpendicular to the quantum well. The polariton group velocity was found to increase with incident power because of polariton scattering.

2.3 Excitonic Polaritons in GaAs Quantum-Well Waveguides: Theoretical Calculations

Experimental results for the delay time of light pulses propagating in GaAs quantum-well waveguides are here quantitatively analyzed on the basis of the first principles calculation of the spatial-dispersion relations [25]. This calculation is based on the additional-boundary-condition-free theory [26] and guided-mode characteristics [45].

Formulation

A model of a waveguide in which the core consists of a quantum well sandwiched between barrier layers is used. The cladding is regarded as a semi-infinite boundary medium. The x axis is taken to be perpendicular to the quantum-well layer and the yz plane is parallel to the layer; the quantum well lies between $z = -L/2$ and $L/2$ in the core, which extends from $z = -d/2$ to $d/2$, where d is assumed to be much larger than L. Figure 7 is a schematic diagram of the quantum-well waveguide.

To take the spatial dispersion of the quantum-well exciton into account without referring to the additional-boundary-condition (ABC) problem, the formulation has to be constructed from the general expression of Maxwell's equations [26]. For the sake of simplicity, only the propagation

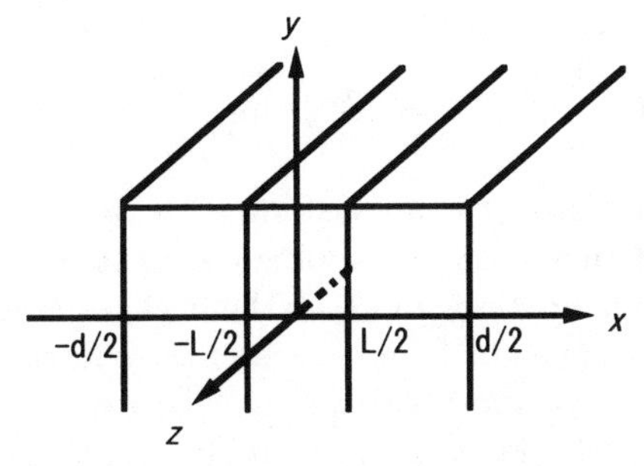

Fig. 7. Schematic coordinates of a quantum-well waveguide

in the TE mode along the quantum-well layer is considered. Here, only the y component of the electric field is nonzero; it may be written in the form $E_y(R) = E(x)\exp(\mathrm{i}k_z - \mathrm{i}\omega t)$, where k_z is a complex propagation constant, ω is angular frequency, and t is time. The amplitude $E(x)$ satisfies the equation

$$\left(\frac{\mathrm{d}^2}{\mathrm{d}x^2} + q^2\right) E(x) + Q^2 \int \mathrm{d}R' \chi(R, R'; \omega) E(x') \exp[\mathrm{i}k_z(z' - z)] = 0 \,, \quad (1)$$

where $q^2 = \varepsilon(\omega/c)^2 - k_z^2$, $Q^2 = 4\pi(\omega/c)^2$, χ is the nonlocal polarizability tensor that is dependent on macroscopic sites $R = (x, y, z)$ and R', which accounts for excitonic excitations and ε is the residual dielectric constant. The tensor indices of χ are omitted, thus assuming a simple situation.

We then consider the energy range around the lowest heavy-hole 1s exciton. Since $L \ll d$, the wavefunction of this exciton is well localized in the z direction within the core. Therefore, the following procedure is applicable to solve (1); we find separate solutions for the core and cladding, and then require that both satisfy Maxwell boundary conditions at the interfaces, where $z = \pm d/2$.

We assume that the energy of the 1s exciton in the core is given by $\hbar\omega_{1\mathrm{s}}(K_{//}) = \hbar\omega_{1\mathrm{s}} + (\hbar^2 K_{//}^2/2M)$, where $K_{//}$ is the wavevector in the yz-plane, $\omega_{1\mathrm{s}}$ is the exciton frequency, and M is the exciton mass. Using the wavefunction given by Matsuura and Shinozuka [46], we now use linear response theory to calculate its polarizability $\chi(R, R'; \omega)$ at low temperatures. Substituting the result into (1), the equation for $E(x)$ in the core is obtained:

$$\frac{\mathrm{d}^2 E(x)}{\mathrm{d}x^2} + q^2 E(x) + Q^2 \chi'(k_z, \omega) \rho^*(x) F = 0 \,, \quad (2)$$

where

$$q^2 = \varepsilon_{\mathrm{b}}(\omega/c)^2 - k_z^2 \,, \quad \chi'(k_z, \omega) = D/(k_z^2 - q_{\mathrm{ex}}^2) \,,$$
$$D = 2M(\mu N)^2/\hbar^2 \,, \quad q_{\mathrm{ex}}^2 = (2M/\hbar)(\omega - \omega_{1\mathrm{s}} + \mathrm{i}\gamma) \,,$$

and

$$F = \int_{-d/2}^{d/2} \mathrm{d}x' \rho(x') E(x') \; , \quad \rho(x) = \cos^2(\pi x/L) \; . \tag{3}$$

In the above equations, μ is a constant with the dimensions of a dipole moment, γ is a phenomenological damping constant, N is the normalization constant of the 1s wavefunction, and ε_b is the background dielectric constant, which includes both the residual constant and the contributions coming from the nonresonant terms of the polarization. The expression $\rho(x)$ is directly related to the form of the 1s wave function [46]. Equation (3) indicates that the 1s exciton is coupled only with the even modes of $E(x)$. Using a method similar to Cho's [26], the solution of (2) is readily obtained as

$$E(x) = C\left(\cos(qx) + \frac{Q^2 D\rho(q) f(q,x)}{\phi(k_z,\omega)}\right) \equiv CG(k_z,\omega;x) \; , \tag{4}$$

where C is a constant and $\phi(k_z,\omega) = k_z^2 - q_\mathrm{ex}^2 - Q^2 Dh(q)$ with

$$\rho(q) = \int_{-d/2}^{d/2} \mathrm{d}x \rho(x)\cos(qx) \; , \quad h(q) = \int_{-d/2}^{d/2} \mathrm{d}x \rho(x) f(q,x) \; , \tag{5}$$

and

$$f(q,x) = -\int_0^x \mathrm{d}x' \frac{\sin[q(x-x')]\rho^*(x')}{q} \; . \tag{6}$$

In (4), the term with the resonant factor $1/\phi(k_z,\omega)$ represents the coupling of light with an exciton. The contribution of this term becomes important in the excitonic resonant region.

The polarization in the cladding is assumed to be local and constant, i.e., $\chi(R,R') = \chi\delta(R,R')$. Therefore, (1) becomes a simple wave equation and its even-mode solution is given by

$$E(x) = C' \exp(\mathrm{i}q'\,|x|) \; , \tag{7}$$

where C' is a constant, $q'^2 = \varepsilon_\mathrm{b}'(\omega/c)^2 - k_z^2$, and $\varepsilon_\mathrm{b}' = \varepsilon + 4\pi\chi$.

The solutions of (4) and (7) must now be required to satisfy Maxwell boundary conditions at the interface $x = d/2$. This leads to

$$\frac{C'}{C} = G\left(k_z,\omega;\frac{d}{2}\right)\exp\left(-\mathrm{i}q'\frac{d}{2}\right) \tag{8}$$

and

$$\mathrm{i}q' G\left(k_z,\omega;\frac{d}{2}\right) - \frac{\mathrm{d}G(k_z,\omega;x=d/2)}{\mathrm{d}x} = 0 \; . \tag{9}$$

The dispersion relation is obtained by solving (9) and the amplitude $E(x)$ is given in terms of C by (7). Ignoring the second term in G of (9), we obtain $\tan(qd/2) = -\mathrm{i}q'/q$, which is a well-known relation in the theory of waveguides.

Numerical Calculation and Comparison with Experimental Results

The same quantum-well structure as was shown in Sect. 2.2 is considered here. The core consists of a 50-Å GaAs quantum well sandwiched between 1.8-μm $Al_{0.3}Ga_{0.7}As$ barrier layers; the cladding is formed by a GaAs layer, and the length of the waveguide (ℓ) is 250 μm. Calculation is restricted to the propagation of the TE mode with energy levels near the heavy-hole 1s exciton. Furthermore, we assume that the barrier layers and cladding are simple media with appropriate dielectric constants. The following parameter values were used in the numerical calculations: $\hbar\omega_{1s} = 1.622$ eV, $\hbar\gamma = 0.035$ meV, $D = 1 \times 10^{-6}$ Å^{-3}, $\varepsilon_b = 12.2$, $\varepsilon'_b = 13.6$, $L/2 = 25$ A, $d/2 = 1.8$ μm. The condition $d \gg L$ assumed in the formulation is thus satisfied.

Equation (9) was numerically solved for real frequencies and many branches were found to appear in the resulting dispersion relation. The real (k_r) and imaginary (k_i) parts of k_z for three of the branches indicated by A, B, and C are plotted in Fig. 8. The other branches show behavior similar to that of branch C, but appear in the higher-energy region of the $\omega - k_r$ plane, and are associated with k_i values larger than those for branch C. Below the resonant region, the dispersions for k_r are photon-like and have a slope of $c/\sqrt{\varepsilon_b}$. In this energy region, only the field corresponding to branch A can propagate along the z axis with a small damping. Approaching the resonant energy, branch A becomes more and more exciton-like. When crossing the resonant region, branches B and C bend over and approach the straight lines that coincide with the extensions of the lower energy parts of branches A and

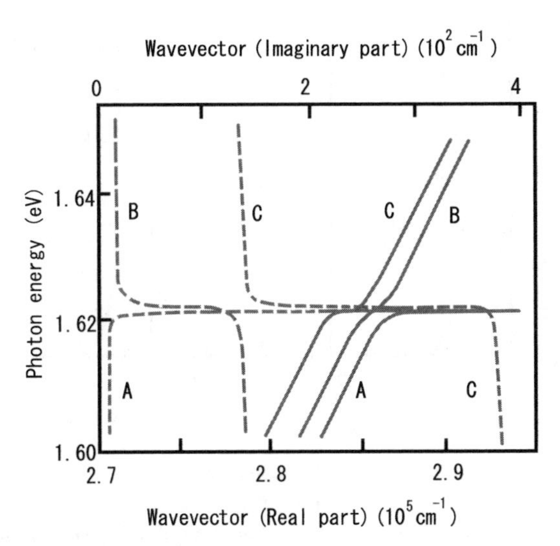

Fig. 8. Calculated dispersion curves. *Solid curves* show the $\omega - k_r$ relation (scale at the bottom) and *dashed curves* show the $\omega - k_i$ relation (scale at the top)

B, respectively. The branch that is associated with small values of k_i changes from A to B on crossing the resonant energy.

Using the field obtained by (8), the z component of the time-averaged Poynting vector, $S(x,z) = (c^2/8\pi\omega)k_\mathrm{r}\,|E_y(x)|^2\exp(-2k_\mathrm{i}z)$ is calculated. Figure 9 shows $S(x, z = 0)$ for branches A and B at different energies; here, only the line shapes can be compared. It is found that branches A and B are the fundamental and the second even mode below the resonant energy, respectively, and when crossing this resonant energy they change their modes. In the same manner, branch C makes a transition from the third to the second even mode on crossing the resonant energy from the low-energy side. At the resonant energy, the amplitude in the quantum well is greatly reduced by the resonant coupling with the quantum-well exciton; especially in branch A, the light power carried by the field has a large amplitude at the interface. This phenomenon is also understood as the result of an enhancement of the effective dielectric constant in the core due to the resonance with the quantum-well exciton.

In the above system, each branch in the dispersion relation along the z direction is characterized in terms of its mode profile, as defined by the dependence of the electric field on x. Furthermore, the branch changes its type of mode depending on the energy because of the coupling with the quantum-well exciton in the resonant region.

From the above results, we find that the excitonic polariton that belongs to the branch associated with the fundamental mode is able to propagate in the z direction without strong damping because of the small value of k_i. The

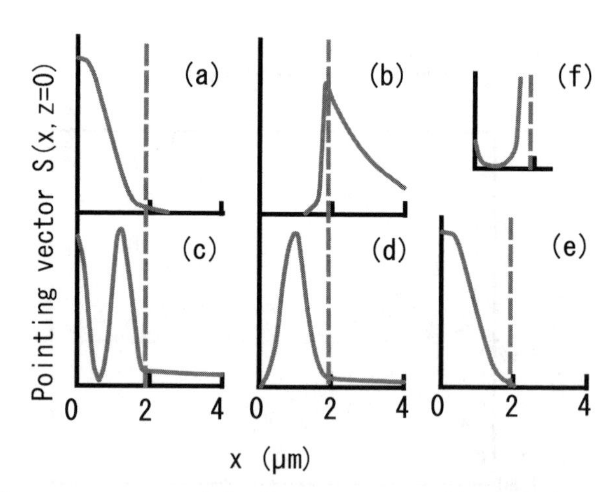

Fig. 9a-f. The x component of the Poynting vector $S(x, z = 0)$ at various levels of energy: (**a**) 1.612, (**b**) 1.6219 eV for branch A, and (**c**) 1.612, (**d**) 1.622, (**e**) 1.632 eV for branch B. (**f**) shows the details of the behavior at around $x = 0$ in (**b**); the ordinate is enlarged by a factor of 50. The *dashed lines* indicate the interface between the core and the cladding. $S(x, z = 0)$ is an even function of x

conclusion from this is that the observed delay time described in Sect. 2.2 is due to the decrease in the group velocity of this polariton. The delay time as predicted by using the group velocity computed from the dispersion relation is plotted in Fig. 10. This result is in reasonably good agreement with the experimental results, which are presented as solid circles.

The absorption observed in the experiment is a difficult quantity to calculate. However, it is reasonable to compare it with the quantity defined by

$$A(\omega) = \ln\left(\frac{\int \mathrm{d}x S(x, z=0)}{\int \mathrm{d}x S(x, z=\ell=250\,\mu\mathrm{m})}\right), \tag{10}$$

which may be regarded as the absorption rate inside the waveguide. The computed $A(\omega)$ is also shown in Fig. 10, where the observed absorption spectrum as defined by $a(\omega)$(absorption coefficient)$\times\ell$ is shown as solid circles; the peak height of $A(\omega)$ is 5.5, while that for $a(\omega)\ell$ is 4.3. The dependence of $A(\omega)$ on ω reproduces the characteristic features of the observed absorption spectrum, although the degree of quantitative agreement is not satisfactory. This is probably because of scattering of the excitonic polariton that is caused by the roughness of the interface between the quantum well and barrier layers and by fluctuations in the composition.

As shown above, the dispersion relation of an excitonic polariton propagating in a waveguide that contains a single quantum well was calculated on the basis of the additional-boundary-condition-free polariton theory [26]. This calculation is concerned with TE modes in the energy region near the heavy-hole exciton state of a GaAs/AlGaAs quantum well. The group velocity and absorption rate obtained from the dispersion relation provide a reasonable explanation of the experimental results.

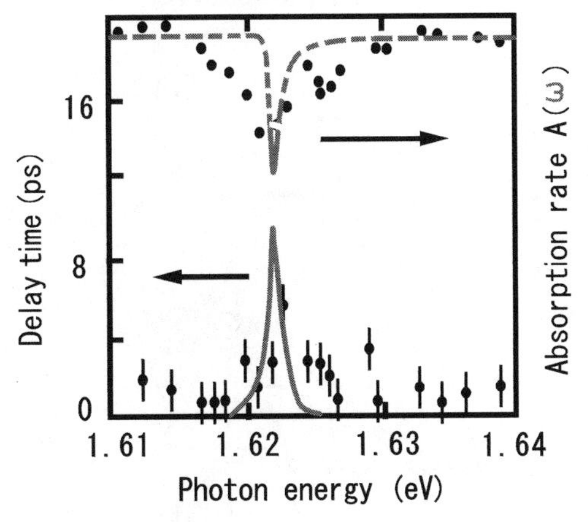

Fig. 10. Calculated delay time (*solid curve*) and absorption (*dashed curve*). *Solid circles* represent the experimental result

2.4 Electric-Field-Induced Phase Modulation of Excitonic Polaritons in Quantum-Well Waveguides

The phase of an excitonic polariton transmitting along a quantum-well waveguide is significantly influenced by applying an electric field. This is because the electric field distorts the exciton wave function and modifies the photon–exciton interaction. Experimental results for such electric-field-induced phase modulation are described below [31–33].

The sample used in the experiment had the same structure as that used for the time-of-flight measurements, except that an electrode for applying an electric field was attached to the waveguide. The substrate of the waveguide is n-type doped GaAs, while the core and cladding layers are undoped. A Schottky-barrier contact is formed on the surface of the cladding to apply a reverse bias to the waveguide. For interferometric measurements, the sample is set in one arm of a Mach–Zehnder interferometer in a liquid-helium cryostat, as is shown in Fig. 11.

Figure 12 is an interference spectrum for a quantum-well excitonic polariton as measured at 6 K. The incident laser beam is linearly polarized in the plane of the quantum well. In this polarization, only the 1s heavy-hole excitons are dipole allowed to show significant effects. Several interference-fringe peaks are seen in the spectral band. In Fig. 13, the photon energies of the five interference-fringe peaks seen around the resonance with the 1s heavy-hole exciton are plotted (solid circles) between 1.607 and 1.615 eV as a function of electric field. For convenience, these peaks are numbered $N = 1, \cdots, 5$ from the fringe at the higher photon energy to that at the lower photon energy.

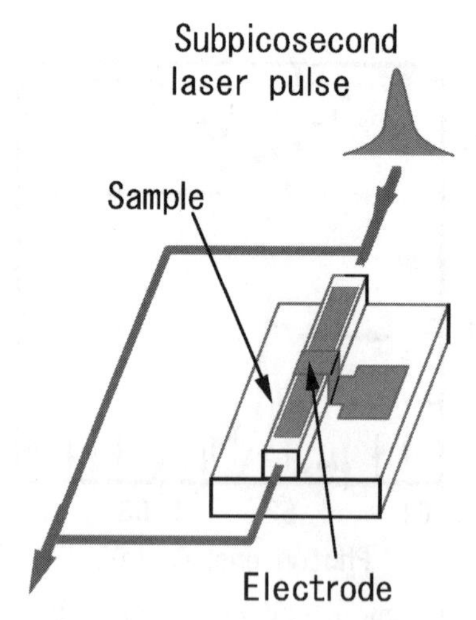

Fig. 11. Schematic view of the setup for interferometry. The sample was placed in one arm of a Mach–Zehnder interferometer in a liquid-helium cryostat at 6 K

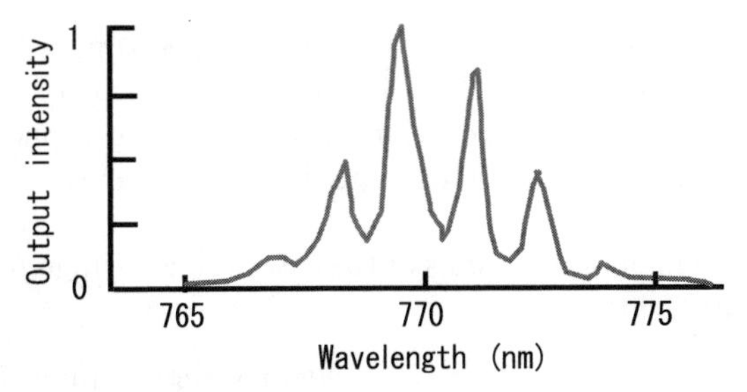

Fig. 12. An example of an interference spectrum for a GaAs quantum-well excitonic polariton. The interference-fringe peaks are seen in the spectral band of subpicosecond laser pulses

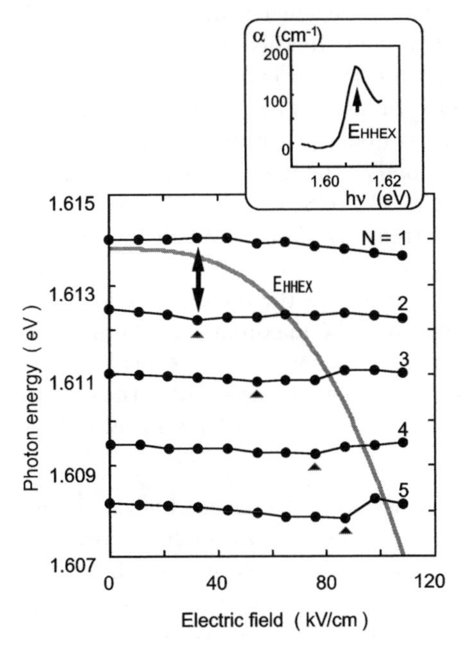

Fig. 13. Dependence at 6 K of the photon energy of interference-fringe peaks near the 1s heavy-hole-exciton resonance on electric field. The peaks are numbered $N =$ 1–5 for convenience. The *solid circles* show experimental data for the fringe peaks and the lines are to guide the eyes. The *gray curve* is an interpolated trace of the electric-field-induced shift of the resonance photon energy of the 1s heavy-hole exciton. The *inset* is the absorption spectrum for 1s heavy-hole excitons in the waveguide structure at 6 K

The resonance photon energy of the 1s heavy-hole exciton is defined as the peak photon energy of the 1s heavy-hole-exciton absorption line. The absorption spectrum around the 1s heavy-hole-exciton resonance under zero electric field is shown in the inset of Fig. 13, where the resonance photon energy is marked with an arrow and labeled E_{HHEX}. The electric-field-induced shift in E_{HHEX} that was measured in the absorption spectrum is interpolated for reference as the gray curve in Fig. 13. The power density of the incident laser is 7.8×10^{-11} J/cm^2 pulse, which corresponds to a photon density of 3.0×10^{10} photons/cm^2 pulse. This value is much lower than the saturation density [24].

The electric-field characteristics of the interference-fringe peaks in Fig. 13 have the following features.

1. The fringe peak $N = 1$ rises to a slightly higher photon energy with increasing electric field until the electric fields reaches $\sim$ 40 kV/cm. As the electric field becomes stronger above this point, the fringe peak shifts to a lower photon energy.
2. The photon energy of each fringe peak of $N = 2, \cdots, 5$ oscillates as a function of the electric field. In this oscillation, the photon energy reaches a minimum under a certain electric field and then rises with further increases in the electric field. The electric field under which the minimum for each fringe peak appears is indicated by a solid triangle. The electric fields that correspond to the minima increase with the order of the peaks, from 33 kV/cm ($N = 2$) to 88 kV/cm ($N = 5$). This shift in the minimum follows the same trend as E_{HHEX}. When the electric field is sufficiently stronger than the level that produces the minima, all of the fringe peaks simply fall monotonically in the direction of lower photon energy, in the same way as seen after the maximum for the fringe peak $N = 1$.
3. The energy spacing between the fringe peaks $N = 1$ and 2 increases with electric field and then decreases after the electric field has exceeded a certain value. The vertical bar with arrows at both ends shows the point of maximum.

The characteristics under an electric field of the fringe peaks $N = 2, \cdots, 5$ are significantly different from those of the fringe peak $N = 1$. This indicates that some physical process that governs the characteristics of the fringes causes a change according to whether the photon energy is higher or lower than E_{HHEX}. This point is later discussed as part of the description of the results on the basis of theory.

The dependence of the time of flight of quantum-well excitonic polariton on electric field was also investigated for fields up to 110 kV/cm. When plotted as a function of photon energy, the time of flight reached a maximum at a photon energy almost equal to E_{HHEX}. The maximum value is about 8 ps under zero electric field, which is consistent with the previous observation as described in Sect. 2.2 [24] and supported by the first-principles calculation as described in Sect. 2.3 [25]. The position of the maximum is shifted in the direction of lower photon energy in accordance with the electric-field dependence of E_{HHEX}. The maximum time of flight is not, however, significantly affected by the low-energy shift. It seems that the strength of the photon–exciton interaction and the translational mass of the 1s exciton remain almost unchanged over the above range of electric field. The dependence on electric field of the interference-fringe and time-of-flight spectra is explainable in terms of ordinary electrorefraction effects [47]. However, the Kramers–Kronig relation shows that the electric-field-induced change that is due to electrorefraction effects is one order of magnitude smaller than the change seen in the experiment. The anomalous dispersion in the refractive index, which always gives a zero time of flight at the peak photon energy of quantum-well exciton absorption, fails to reproduce the time-of-flight spectrum [24]. Electrorefraction effects are thus neglected in the theoretical treatment below.

The theoretical basis for the dependence of the positions of the interference-fringe peaks on electric field is given here by a phenomenological model for quantum-well excitonic polaritons. The phenomenological model embodies the unique characteristics of the photon–exciton interaction in the waveguide structure. An effective LT splitting energy of two-dimensional excitons in the waveguide structure, $\Delta'_{2\mathrm{DLT}}$, is employed instead of the conventional LT splitting energy of two-dimensional excitons, $\Delta_{2\mathrm{DLT}}$. The effective splitting energy is defined in terms of the conventional splitting energy as $\Delta'_{2\mathrm{DLT}} = \eta \Delta_{2\mathrm{DLT}}$, where η is the mode-confinement factor (filling factor) [47], which expresses the proportion of the light power within the quantum well of all light power within the core. For the above sample, η was found to be 7.8×10^{-4}. Light-mode characteristics in the waveguide structure were also included in the model. The spatial-dispersion curves calculated from the model are shown in Fig. 14. The black curves are spatial-dispersion curves under zero electric field, while the gray curves are for dispersion under an electric field of 12 kV/cm. The curve PB1 represents the exciton-like polariton lower

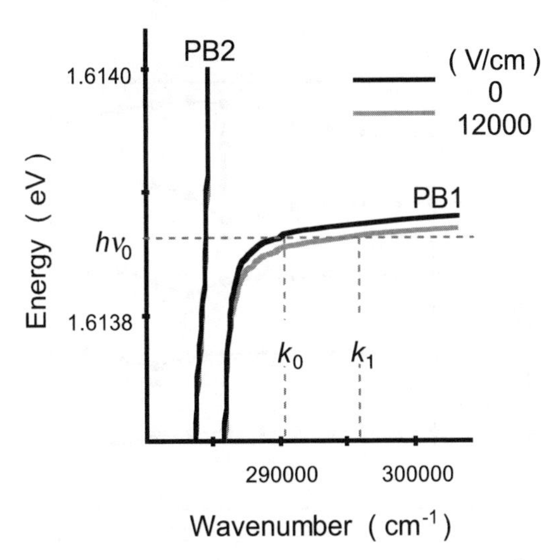

Fig. 14. Spatial-dispersion curves for polaritons as calculated from a phenomenological model. The *black curves* are dispersion curves under zero electric field and the *gray curves* are for dispersion under an electric field of 12 kV/cm

branch, and the curve PB2 the photon-like polariton upper branch. These curves coincide well with spatial-dispersion curves given by first-principles calculation [25]. The branches PB1 and PB2, respectively, correspond to the first and second guided modes for quantum-well excitonic polaritons that were obtained in this calculation. The shift towards low energy of the energy level of the quantum-well exciton is included in this phenomenological model, while the strength of the photon–exciton interaction and the translational mass are kept constant. The dependence on electric field produced by the model results in a shift toward low energy of the spatial-dispersion curves. This redshift induces a change from k_0 to k_1 in the wavenumber at a fixed photon energy $h\nu_0$ as is shown in Fig. 14, and leads to an optical phase shift according to the relation $\Delta\psi = (k_1 - k_0)\Delta L$. Here, ΔL is the distance along which the quantum-well excitonic polaritons experience phase modulation in the electric field. This phase shift of the polaritons produces the dependence of the interference spectrum on electric field.

The peak photon energy (solid circles) of the interference fringes as given by the calculated spatial-dispersion curves is plotted against the electric field in Fig. 15. The calculated peak photon energy of the fringes is given in electric-field increments of 11 kV/cm, the interval used in the experiments (Fig. 13). Each fringe peak is numbered in the same manner, too. The calculated fringe peaks $N = 1, \cdots, 5$ qualitatively reproduce the features of the electric-field characteristics of the corresponding experimental fringe peaks. The calculated fringe peaks $N = 2, \cdots, 5$ also show oscillation as a function of electric field, and the electric field where the minimum occurs moves with the fringe peaks in the same way as was seen in the field dependence of

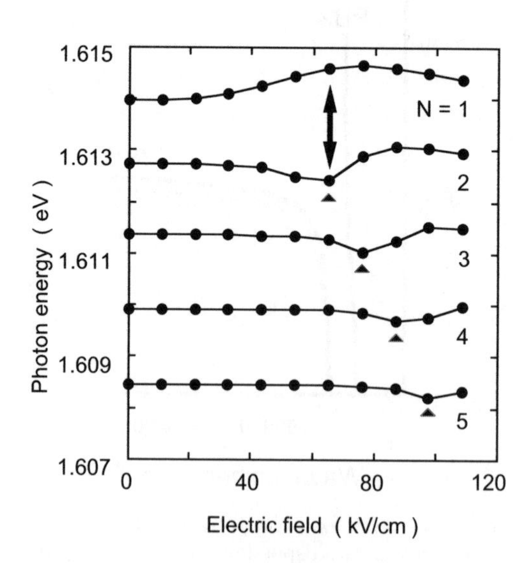

Fig. 15. Calculated fringe-peak photon energy vs. electric field

E_{HHEX}. The energy spacing between the fringe peaks $N = 1$ and 2 reaches its maximum at the minimum of $N = 2$. By fitting the experimental data to the theoretical dispersion relation with the conventional LT splitting energy (Δ_{2DLT}) as a parameter of fit, a value of 0.4 meV was obtained for Δ_{2DLT}. This is five times larger than the LT splitting energy of 1s excitons in bulk GaAs (0.08 meV) [9]. This increase in the LT splitting energy characterizes the quantum-confinement effect on excitonic polaritons in the waveguide structure.

Applying an infinite-confinement model to the quantum-well exciton provides the following qualitative explanation of the above increase. Within the framework of the dipole approximation, this quantum-confinement effect is caused by an increase in the 1s oscillator strength of the quantum-well exciton. Theoretically, an increase in the 1s oscillator strength would be expected to increase by eight times the increase for a bulk exciton, because the oscillator strength of the n-th s-state exciton in two dimensions is proportional to $1/(n-1/2)^3$ instead of the $1/n^3$, which applies in the three-dimensional case $(n = 1, 2, \cdots, \infty)$ [48]. This leads to an eight-fold increase in the LT splitting energy of quantum-well excitonic polaritons according to the following relation [6]:

$$\Delta_{\mathrm{2DLT}} = \frac{2\pi\hbar^2 q^2 f_{\mathrm{QWEX}}}{\kappa \mu_{\mathrm{ex}} E_{\mathrm{ex}} V_{\mathrm{ex}}} \,, \tag{11}$$

where f_{QWEX} is the 1s oscillator strength, κ is the dielectric constant, μ_{ex} is the exciton reduced mass, and E_{ex} ($= E_{\mathrm{HHEX}}$) is the transverse exciton energy. The volume of the quantum-well exciton is described as

$$V_{\mathrm{ex}} = \pi a_{\mathrm{QWEX}}^2 L_{\mathrm{QW}} \,. \tag{12}$$

The radius of a quantum-well exciton $a_{\mathrm{QWEX}}$is estimated from a variational calculation [49]. L_{QW} is the quantum-well width. Substituting $\Delta_{\mathrm{2DLT}} = 0.4$ meV and $E_{\mathrm{HHEX}} \approx 1.62$ eV into (11), the 1s-state oscillator strength f_{QWEX} is calculated to be 0.4 on the basis of reported values, $\kappa = 13$, $\mu_{\mathrm{ex}} = 0.041 m_0$, $E_{\mathrm{ex}} = E_{\mathrm{HHEX}} \approx 1.62$ eV, and $a_{\mathrm{QWEX}} \approx 0.5 a_{\mathrm{B}} = 6.8$ nm [9,49]. The 1s quantum-well exciton carries a large proportion, about 40%, of the total oscillator strength.

We now discuss the physical process that produces the oscillation with electric field seen in Figs. 13 and 15 on the basis of a transfer of the branch of the quantum-well excitonic polariton that governs the photon–exciton interaction. In an energy region that is sufficiently far below the longitudinal exciton energy (i.e., $\ll E_{\mathrm{HHEX}} + \Delta'_{\mathrm{2DLT}}$) the quantum-well excitonic polaritons associated with the branch PB1 are dominant in interactions with the incident photons. The propagation of the light pulses is governed by the polariton branch PB1 in Fig. 14. When an incident photon has an energy that approaches the longitudinal exciton energy, dominance over the interaction is transferred from branch PB1 to branch PB2. A negative phase shift takes place in this transfer from PB1 to PB2. This negative phase shift makes the fringe peaks jump into a higher-energy region. On the other hand, in an energy region sufficiently far above the longitudinal exciton energy, the branch PB2 is dominant in the interaction with the incident photons. Since the energy of the longitudinal exciton energy shows a low-energy shift with increasing electric field, the polariton branch effective in the interaction transfers from PB1 to PB2 with increasing electric field. Such a transfer is only seen for those fringe peaks whose photon energy is well below the longitudinal exciton energy. Since the shift of the longitudinal exciton energy is dominated by the shift of E_{HHEX}, the position of the maximum also moves in the direction of higher electric field from the fringe peak $N = 2, \cdots, 5$, as is shown in Figs. 13 and 15.

On the other hand, it is very important to measure the dependence of the phase modulation of these excitonic polaritons on light intensity because the existence of the optical nonlinearity of the polariton can be examined. First, the absorption spectra of the waveguide at incident photon densities that were either LOW (3.0×10^8 photons/cm^2 pulse) or HIGH (6.2×10^{10} photons/cm^2 pulse) were measured, as shown in Fig. 16. When the electric field is zero, the saturation of the heavy-hole (hh) exciton absorption (1.615 eV) at HIGH density is not prominent. Under a field of 43 kV/cm, on the other hand, nearly 10% of the saturation of hh-exciton absorption is seen in the HIGH photon density. The electric field decreases the exciton binding energy and increases the exciton Bohr radius. A larger Bohr radius lowers the saturation density, and the saturation of the hh-exciton absorption occurs under the electric field. This hh-exciton saturation reduces the oscillator strength of the quantum-well excitonic polaritons.

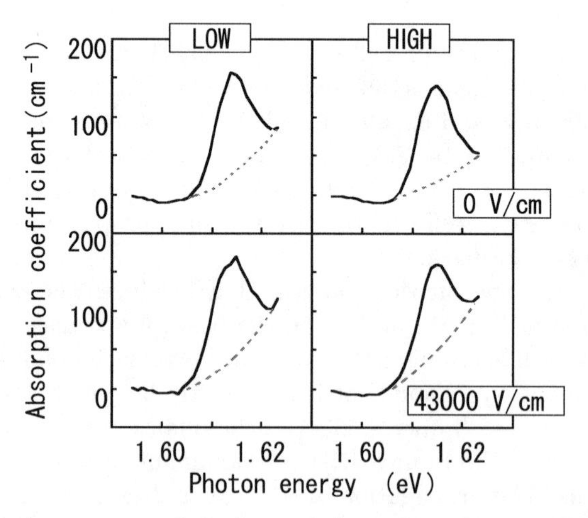

Fig. 16. Absorption spectra of the heavy-hole excitons in a waveguide at 'LOW' and 'HIGH' incident photon densities (3.0×10^8 and 6.2×10^{10} photons/cm^2 pulse)

The dependence of phase change of the polariton propagation on electric field changes markedly with the incident photon density, as is shown in Fig. 17. The photon energy in this case is 1.611 eV, which is slightly lower than the hh-exciton absorption peak. This dependence of the field-induced change on photon density shows the existence of a strong optical nonlinearity that is not explainable in terms of the reduction of the oscillator strength that is suggested by the absorption saturation: the phase shift at the HIGH density is much smaller than that expected as a result of the reduced oscillator strength. In Sect. 2.2, the decrease in the time of flight at high photon density is explained by introducing a pronounced polariton–polariton scattering. We thus expect that such polariton–polariton scattering strongly affects the saturation behavior of the field-induced phase change.

In summary, the electric-field dependence of the propagation of quantum-well excitonic polaritons in a waveguide structure was investigated by interferometric spectroscopy. By using a phenomenological model for the polaritons, the dependence of interference fringes on electric field was calculated. The model qualitatively supports the experimental observation by taking into account the lower-energy shift of the dispersion curves of the quantum-well excitonic polariton. The LT splitting energy of two-dimensional excitons in a GaAs quantum well is 0.4 meV, five times larger than the splitting energy of three-dimensional excitons in bulk GaAs. This reflects the quantum-confinement effect on excitonic polaritons. The increased LT splitting energy corresponds to the oscillator strength of a 1s-state quantum-well exciton as

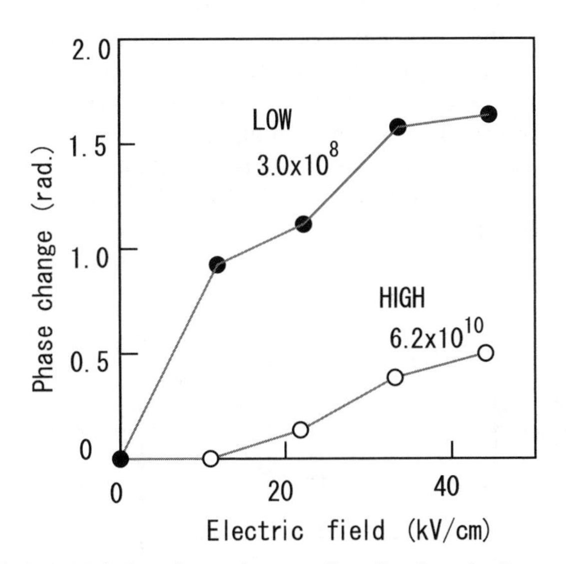

Fig. 17. Dependence of the phase change of excitonic-polariton propagation on electric field. The photon energy is 1.611 eV. The 'LOW' and 'HIGH' incident photon densities are 3.0×10^8 and 6.2×10^{10} photons/cm^2 pulse, respectively

large as 0.4. This means that the 1s-state quantum-well exciton carries 40% of the total exciton oscillator strength.

The dependence on light intensity of the propagation of excitonic polaritons in a quantum-well waveguide was also investigated. Optical nonlinearity in the field-induced phase shift of the excitonic polariton was observed. The saturation properties of quantum-well excitons are unable to account for this dependence and this suggests that polariton–polariton scattering plays an important role.

2.5 Temperature Dependence of the Phase Modulation due to an Electric Field

In Sect. 2.4, we described the phase change of the excitonic polariton transmitted along the GaAs quantum-well waveguide at low temperatures, e.g., in the liquid He temperature range. However, if we wish to apply this aspect of polariton propagation in optoelectronic devices, we usually require operation at much higher temperatures, such as at room temperature. Therefore, it is important to study the phase change at much higher temperatures. In this section, we describe the experimental results on the temperature dependence of these phase changes at much higher temperatures [34,38].

In order to study this temperature dependence, we used a sample that consists of a 2-µm-wide GaAs/AlGaAs ridge-type waveguide and a 400-µm-long electrode to apply the electric field. Figure 18 shows the cross-sectional structure of the waveguide, which has a p-i-n structure of GaAs/AlGaAs layers. A 1.8-µm-thick $Al_{0.13}Ga_{0.87}As$ core layer is sandwiched between $Al_{0.17}Ga_{0.83}As$ cladding layers. At the center of the core layer, a 7.5-nm GaAs single-quantum well is formed. The resonance wavelength of the quantum well is 790 nm. This waveguide supports no higher mode, thus it is single-mode. The light propagating in the waveguide is well confined and interacts with excitons in the quantum well, forming stable polariton propagation. Modal calculation yields a mode confinement factor of $\eta = 4.5 \times 10^{-3}$. This is defined as the overlap between the quantum wells and the optical mode.

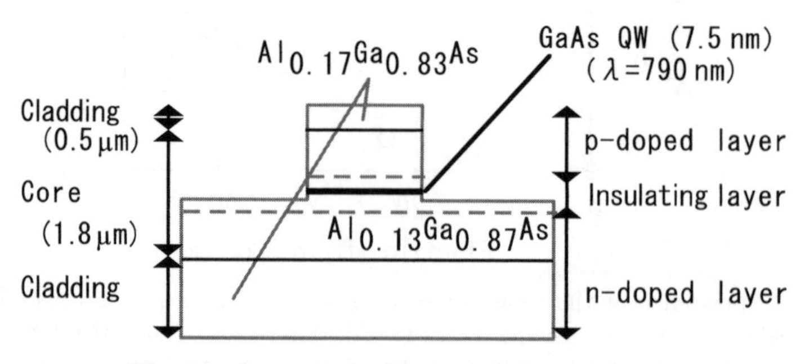

Fig. 18. Cross-sectional view of the waveguide

A tunable Ti-sapphire laser, excited by an Ar laser, was used as a light source. The average power used for this measurement was 20 mW. To obtain the interferometric measurements, the sample and a separate reference path were arranged to form a Mach–Zehnder interferometer. The TE-polarized beam was focused on the end face of a sample in a cryostat, which was inserted in one arm of the Mach–Zehnder interferometer. The interference pattern was observed with a CCD camera. The temperature of the sample was controlled by a temperature controller and was varied through the range from liquid-helium temperature (LHeT) to room temperature (RT). The phase change due to the applied electric field was deduced from the resulting change in the fringe pattern.

Figure 19 shows the relation between observed phase shift per unit voltage and the amount of detuning from the resonant wavelength. The phase-shift rate, which is directly related to the refractive index change Δn, decreases monotonically with increase in the detuning wavelength. Note that the dependences of the phase-shift rate on detuning are classified into two groups, one corresponding to results for lower temperatures (10–80 K) and the other for higher temperatures (160–286 K).

If we focus our attention on the temperature dependence, we find that the phase-shift rate is considerably larger in the lower than in the higher temperature range. For example, when the measured wavelength is detuned by 10 nm from the resonant wavelength, the phase shift seen below 80 K is around π per unit voltage. On the other hand, at temperatures above 160 K, the shift is about $\pi/10$ per unit voltage. At the higher temperatures, the rate of change in the refractive index per unit electric field ($\Delta n/n/E$) is about

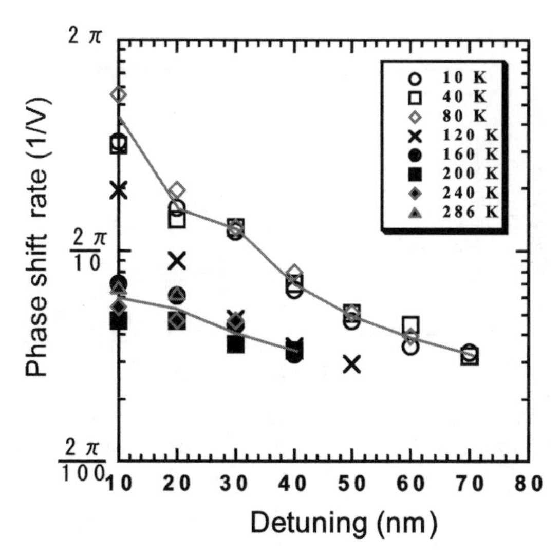

Fig. 19. Relation between the observed phase shift per unit voltage and amount of detuning wavelength

4×10^{-10} cm/V when $\eta = 4.5 \times 10^{-3}$. This result is roughly in agreement with the result on normal light propagation [50]. The critical temperature is around 120 K. This temperature dependence is shown more clearly in Fig. 20, which shows the phase-change rate at a detuning of 10 nm as a function of temperature.

Thus, our measurements showed that the large phase shifts can be obtained at least up to 120 K. On the other hand, as was discussed in the introduction, the LT splitting energy defines the polariton stability with temperature. The GaAs quantum well gives the LT splitting energy of 0.4 meV [30], which implies that a well-defined polariton only exists up to 4.6 K. Accordingly, the critical temperature of 120 K obtained in the experiment is remarkably higher than had been expected on the basis of this criterion. Thus, the observed large critical temperature is not simply explainable in terms of the polariton stability as defined by the LT splitting energy. This indicates that the real part of the dielectric constant is influenced in some complex manner by the polariton effect. Further theoretical and experimental investigation is required to clarify the physical origin of this critical temperature of 120 K.

In summary, we investigated the temperature dependence of the polariton phase-change under an electric field, using interferometric measurement. The phase-change rate is considerably larger in the lower than in the higher temperature range. The large phase-change rate remains up to 120 K. This critical temperature is much higher than we had expected, and the physical mechanism responsible is not clear at present. Such a high critical temperature indicates the possibility of the operation of polariton-based devices at relatively high temperatures.

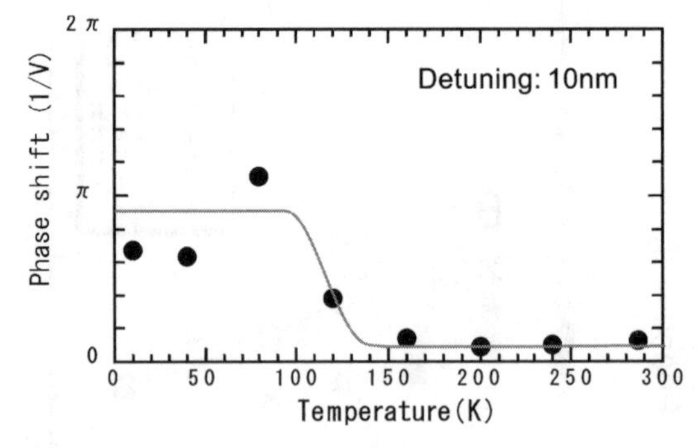

Fig. 20. Temperature dependence of the phase change of the output light from the waveguide for a detuning of 10 nm. The *solid line* is to guide the eye

2.6 Cavity Effect of Excitonic Polaritons in Quantum-Well Waveguides

Enhancement of the coupling between the exciton and photon is particularly important if the excitonic polariton is to be applied in actual optoelectronic devices. In recent years, such enhancement has been widely discussed by using a semiconductor vertical cavity [51], which also plays an important role in the analysis of phenomena such as laser operation and quantum fluctuation. The strong coupling of the excitons in the quantum well with the vertical cavity modes can lead to a splitting in the exciton luminescence spectra, which corresponds to vacuum Rabi splitting. The magnitude of this splitting may be estimated by a simple model based on the dispersion relation of the polariton [52].

By analogy with such a vertical cavity polariton, the cavity effect on a polariton being transmitted along a waveguide is effective in improving the coupling between the exciton and photon. In this section, we report on our experimental results concerning the luminescence based on two cases of the propagation of an excitonic polariton in a GaAs quantum-well waveguide: with and without grating waveguides. We discuss the differences between the luminescence spectra for the two types of samples, and explain why we think that the fine structures in the spectra originate in 'splitting' that is due to the exciton–photon interaction. We were able to obtain a seven-fold increase in interaction energy by using the gratings in the waveguide structures compared to the original energy, and found that control of the wavenumber of the polariton is important to achieve this increase.

In most experiments, double-peak structures in the photoluminescence (PL) spectra are used to indicate the magnitude of the LT splitting of the materials. The LT splitting refers to the difference between the energy of the longitudinal exciton at $k = 0$ and the energy of the transverse exciton at $k = 0$. The amount of the LT splitting is indicated in Fig. 21. If the momentum (which is equivalent to the wavevector) of the polariton is fixed within a narrow region, then the interaction energy of the system is described by

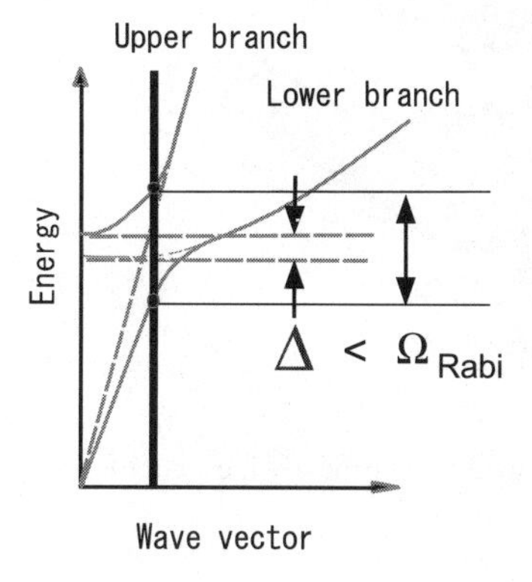

Fig. 21. Dispersion relation of an excitonic polariton. The Rabi splitting energy Ω_{Rabi} is much larger than the LT splitting energy Δ

a parameter called the Rabi splitting. This situation is schematically illustrated in Fig. 21. As shown in the figure, Rabi splitting is much larger than LT splitting, so we can enlarge the interaction beyond the photon and exciton broadening energy.

If phonon scattering is so weak that energy loss is negligible, then polaritons should propagate with fixed momentum in the waveguide structures. In actual samples, since dissipation caused by phonons cannot be neglected, we cannot observe Rabi splitting in the PL spectra of a quantum-well waveguide. On the other hand, in a waveguide with a grating that has a 1/4-wavelength shift, the momentum of the polariton is restricted to a narrow region according to the transmission spectrum of the waveguide. In other words, when the momentum of the polariton is reduced by interaction with the phonon, the intensity of the polariton is rapidly decreased. We fabricated a grating such that its Bragg wavelength matched the wavelength of the radiation from the excitons. The difference between these two wavelengths must coincide precisely with the magnitude of Rabi splitting.

A schematic cross-sectional view of the waveguide sample used in the experiment is shown in Fig. 22. An important feature of the waveguide is that a single GaAs quantum-well layer (7.5 nm thick) occupies the center of the cross section of the waveguide. This quantum-well layer is sandwiched between AlGaAs/GaAs superlattice layers, and these layers together form the light-transmitting core region of the waveguide. The interval of the grating was 225 nm. The grating was fabricated by electron-beam (EB) lithography and chemical etching and was parallel to the ⟨010⟩ direction. The depth of the grating was about 0.1 μm. The grating's Bragg wavelength was tuned to the wavelength of exciton radiation by slightly rotating the waveguide relative

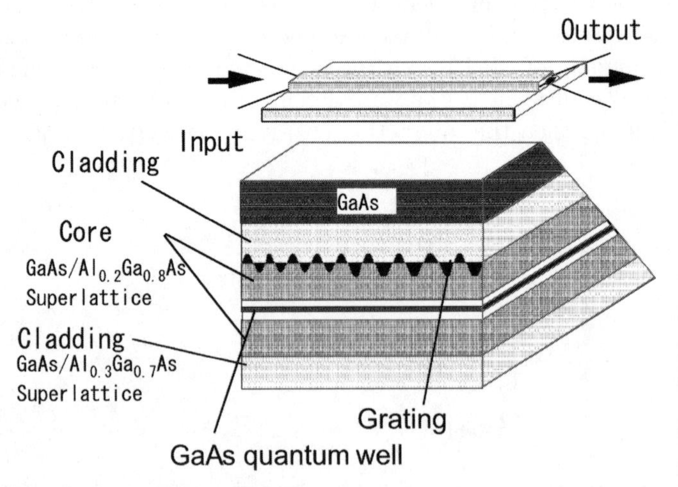

Fig. 22. Cross-sectional view of the quantum-well waveguide with grating

to the grating direction. We also fabricated a sample without a grating, but otherwise identical, for use in comparison.

We focused the excitation light on one end-face of the sample and observed the PL spectra of the light emitted from the other end-face after passing through the quantum-well waveguide. The sample was kept at 4.2 K. The excitation energy was directed to the transparent part of the core region by using a Ti:sapphire laser with an Ar-laser pump.

In describing the experiment with the samples that had gratings, we refer to coincidence of the interval of the grating with the wavelength of the exciton radiation as on-resonance. In this case, we obtained spectra with double peaks. An example is shown in Fig. 23. The split between the peaks is about 3.1 meV wide. In the off-resonance situation, on the other hand, we obtained narrow single-peak spectra that correspond to heavy-hole-exciton resonance. We examined the excitation power and temperature dependence of the PL spectra and transmission spectra of the waveguides under the on-resonant condition (Fig. 24) to ensure that the double-peak structure was not being caused by reabsorption or the effects of impurities on luminescence. As the excitation intensity is increased, the double peak becomes less prominent. For high-power excitation, i.e., levels beyond 50 mW, the double peak disappears. This is consistent with our belief that a high density of excitons weakens the oscillator strength [53]. Furthermore, raising the temperature of the system to 60 K also eliminates the double-peak structures in the spectra. This indicates that the interaction energy is less than 5.2 meV. Thus, we feel confident that the double-peak structure corresponds to Rabi splitting.

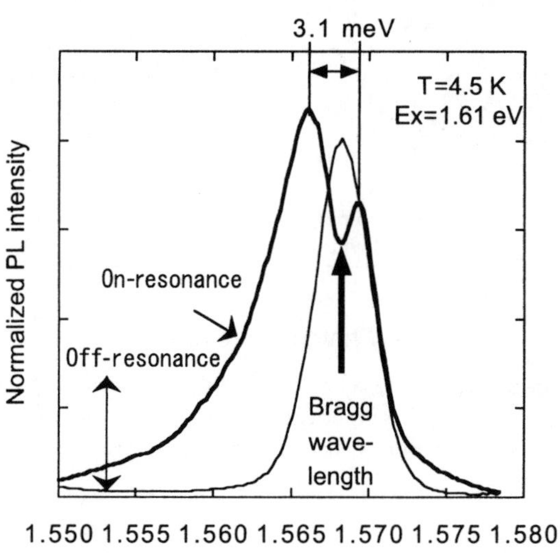

Fig. 23. Photoluminescence spectra of a quantum-well waveguide. Excitation energy is 1.61 eV. Measured temperature is 4.5 K

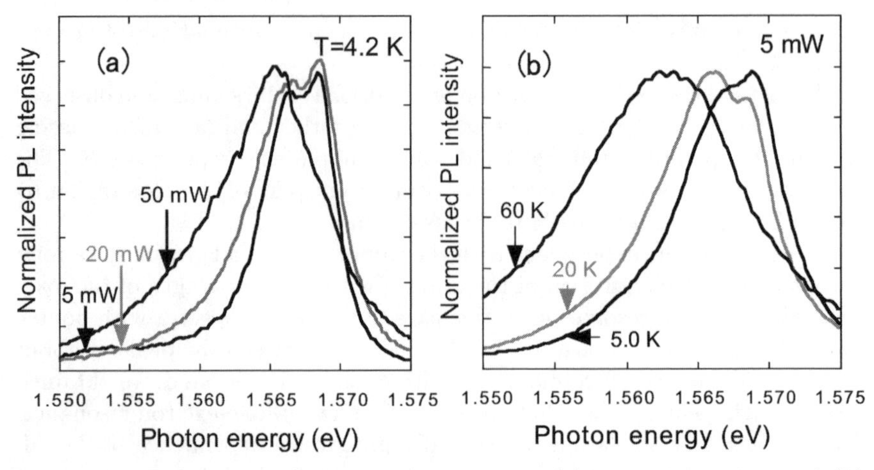

Fig. 24a,b. Dependence of photoluminescence on (**a**) excitation power and (**b**) temperature for 'on-resonant' samples. In (**a**), the measured temperature is 4.2 K, and the excitation powers are 5, 20, and 50 mW, respectively. In (**b**), the excitation power is 5 mW, and the measured temperatures are 5.0, 20, and 60 K, respectively

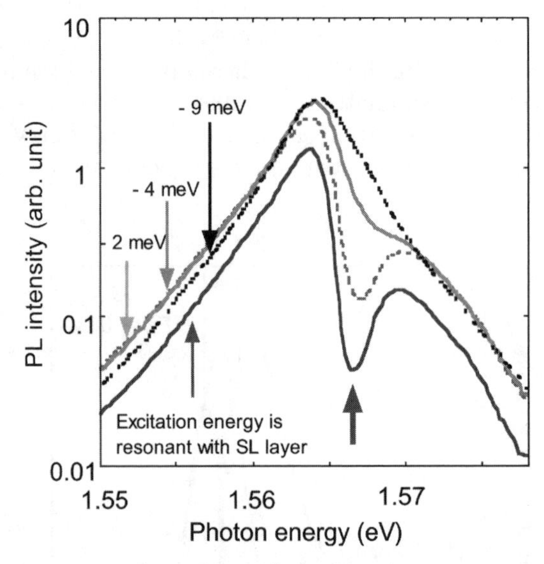

Fig. 25. Excitation-energy dependence of photoluminescence spectra for a sample with no grating. The measured temperature is 10 K. The resonant excitation energy is 1.66 eV. The values in the figure correspond to amounts of energy shift from the resonant energy

We also obtained a narrow transmission peak that is based on the Bragg condition of the grating. This peak is about 0.6 meV wide, too narrow to be attributed to the superposition of the broad luminescence peak and the narrow transmission peak, because each of the luminescence peaks in the double-peak spectra is a few meV wide.

The vacuum Rabi splitting can be quantum mechanically described in the following way [52]

$$\Omega_{\mathrm{Rabi}} = \frac{2d}{\hbar}\sqrt{\frac{\hbar\omega_{\mathrm{g}}}{\varepsilon_0 V_{\mathrm{cav}}}} = 2\sqrt{\frac{\Delta\cdot\omega_{\mathrm{g}}}{\hbar}\frac{V_{\mathrm{ex}}}{V_{\mathrm{cav}}}}\,, \tag{13}$$

where d is the dipole moment of the QW (quantum-well) layer, ε_0 is the dielectric constant of the core layer, ω_{g} is the circular frequency of the band-gap, Δ is the energy of LT splitting, V_{ex} is the volume of QW, and V_{cav} is the effective volume of a cavity. When we applied this formula to our structures, we were able to estimate that Rabi splitting was 3.2 meV for an LT splitting of 0.4 meV [32]. This value is consistent with our spectral results. The energy difference between the two peaks does not directly indicate vacuum Rabi splitting, but the coupling coefficient for our sample was 65 cm^{-1}. This is so small that the deviation from the true value of Rabi splitting can be ignored.

Next, we discuss the PL spectra obtained from the sample without the grating. We again observed double-peak structures in the PL spectra from the edge of the QW waveguide when the energy of the excitation light was higher than the resonance energy of the core layers (i.e., the superlattice layer) of the waveguides. Figure 25 shows PL peaks that correspond to radiation from heavy-hole excitons. We have two reasons to believe that the dip structure seen here is caused by reabsorption due to resonance with the heavy-hole exciton. First, our measurements of transmission spectra indicated that the absorption peak energy due to heavy-hole excitons in the transmission spectra coincides with the dips seen in the PL spectra. Secondly, these structures were still visible in results for high-power excitation and high temperatures of measurement, up to 80 K. Thus, we believe that the reason for the non-coincidence of the peak and dip positions in these spectra is the same as the reason for the Stokes shift. For the sample with the grating, the dip in the spectrum appeared at the center of the peak. This fact indicates that the double-peak structure cannot be regarded as a peak and a dip.

In conclusion, we examined the validity of using waveguide structures with gratings to obtain strong coupling between excitons and photons by applying the same principle as is used in the vertical cavity. The interaction energy is 3.1 meV in these structures. This is consistent with the results of calculation using a model based on the dispersion relation of the polariton. This value is seven times larger than the magnitude of LT splitting. Double-peak structures were thus visible in the PL spectra because of coherent exciton–photon interaction at relatively high temperatures.

3 Applications to Optoelectronic Devices

As was discussed in Sect. 1, the propagation of excitonic polaritons in waveguides provides a way to realize advanced optical devices. The low-voltage operation of devices based on polariton propagation would be particularly useful in the present optical communications network systems, because low-voltage operation leads to low levels of power dissipation and high operational drivability. These factors become much more important with increasing transmission capacity. In the following sections (Sects. 3.1 and 3.2), the application of the excitonic polariton effect in two typical devices, i.e., Mach–Zehnder-type modulators and directional-coupler-type switches, is described with a focus on low-voltage operation.

Furthermore, the reduction in the size of optical devices that is offered by the polariton effect presents the possibility of such advanced optical systems as optical interconnections in LSI chips and small circuit boards. Reducing the sizes of individual devices to the nanometer level is discussed in Sects. 3.3 and 3.4.

3.1 Mach–Zehnder-Type Modulators

The operational principle of a Mach–Zehnder-type modulator is interference between the waves being transmitted in a pair of separated waveguides. Changing the phase of the wave transmitted in one of the waveguides modulates the output power. The phase-change is induced by an applied electric field.

Figure 26a is a schematic illustration of the Mach–Zehnder (MZ) modulator, which consists of a GaAs/AlGaAs single-mode ridge-type waveguides [38,54,55]. The cross-sectional structure of the waveguide is shown in Fig. 26b. A 1.8-μm-thick core layer of $Al_{0.13}Ga_{0.87}As$ was sandwiched between $Al_{0.17}Ga_{0.83}As$ cladding layers. Ten pairs of 7.5-nm GaAs/$Al_{0.3}Ga_{0.7}As$ quantum wells were formed at the center of the core layer. The resonance wavelength of these quantum wells was 803 nm at 4.5 K. The waveguide has the same structure as that described in Sect. 2.5. The 1.06-mm long electrode

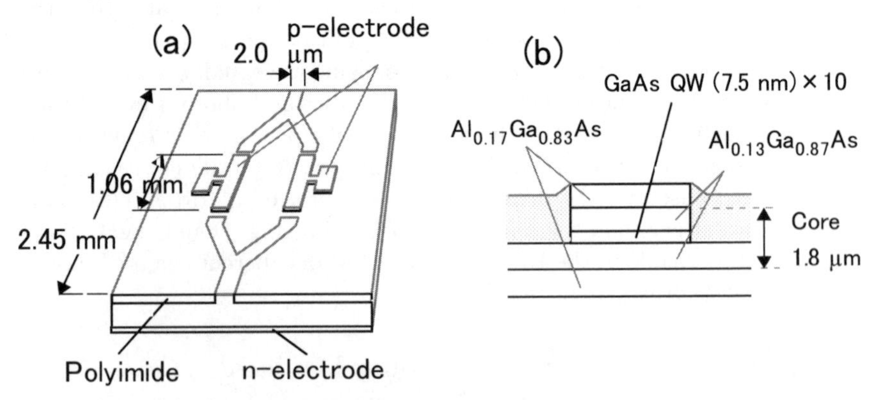

Fig. 26a,b. Schematic illustrations of a Mach–Zehnder-type modulator that uses polariton propagation; (**a**) an overview and (**b**) a cross section of the waveguide

was formed on the straight regions of the waveguides. The total length of the device was 2.45 mm. Figure 27a is a plan view of the actual device, which has the shielding electrodes that are necessary for stable operation. Figure 27b is an SEM image of the device.

The output power is modulated by an applied electric field. Figure 28a shows the experimental result for output intensity as a function of applied voltage. These measurements were made 4.5 K. The measuring wavelength is detuned by 12 nm from the resonant energy of the quantum well. As is shown in the figure, the output power decreases as the applied voltage is increased until it reaches the minimum power of −11 dB. This minimum power is obtained by applying only 0.7 V. This voltage is quite low compared with those demonstrated in conventional optical modulators, thus the device achieved low-voltage operation by the application of polariton propagation. Note that the power recovers to the initial level after passing the minimum output power. This means that the transmission loss in relation to the exciton absorption does not affect the output power, although the measured wavelength is only detuned by 12 nm.

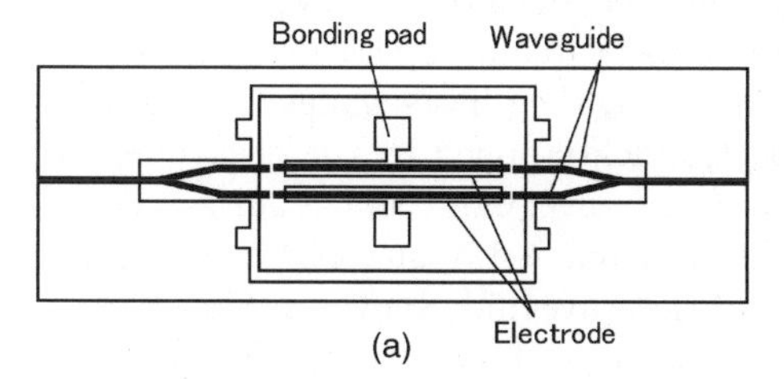

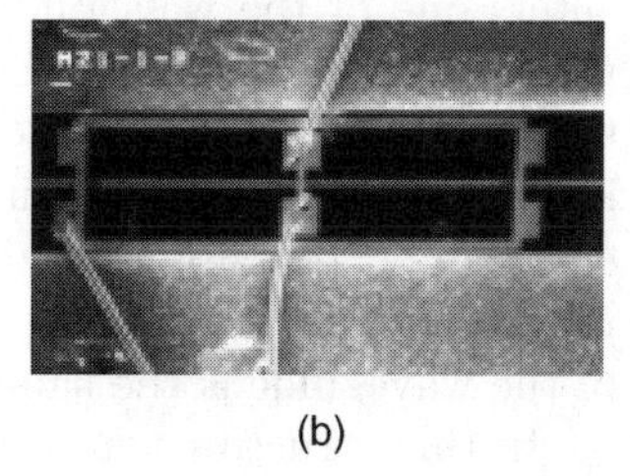

Fig. 27a,b. Fabricated Mach–Zehnder-type modulator: (**a**) plan view and (**b**) scanning-electron-microscope image

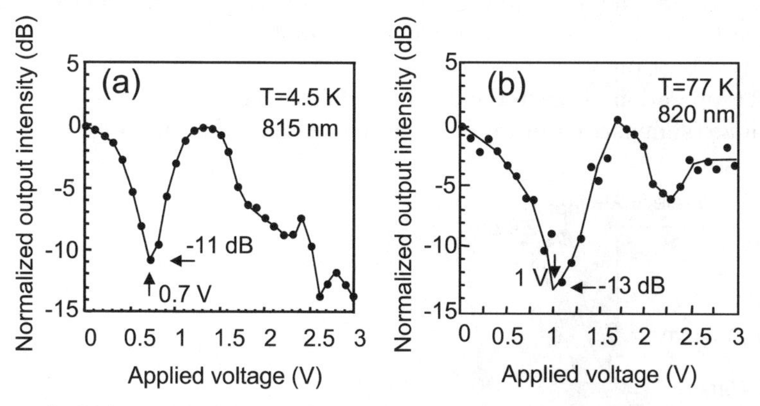

Fig. 28a,b. Voltage dependence of normalized output-light intensity for the Mach–Zehnder-type modulator at (**a**) 4.5 K and (**b**) 77 K. The detuning wavelength was the same in both cases

The operation characteristics at much higher temperatures were also measured. Figure 28b shows the normalized output intensity as a function of the applied voltage. The wavelength was detuned by the same amount as had been used at 4.5 K, and similar results are obtained. The voltage that gives the minimum output power is still low, at 1 V, and the minimum power is now −13 dB below the initial power. Low-voltage operation is thus obtained in the range up to at least 77 K. This characteristic coincides with the observed temperature dependence of phase-change under an electric field, which was presented in Sect. 2.5.

3.2 Directional-Coupler-Type Switches

A directional-coupler-type switch is another typical device to which a phase change under an electric field is applied [37,38]. The device we fabricated is schematically illustrated in Fig. 29. This device consists of a couple of 2-μm wide waveguides and an electrode to apply the electric field. As is shown in the figure, the two waveguides approach each other in the central coupling region. The distance between the waveguides is typically 1 μm. The light beam enters one of the waveguides and is switched to another waveguide by the phase-change in this coupling region. This switching operation is based on the change in the mode field that is caused by the phase-change. The switching is controlled by adjusting the electric field applied by the electrode. An electrode is actually formed on each of the waveguides, but a voltage is applied to one of them. The length of the electrode is 1.3 mm. The cross section of the waveguide is the same as that of the waveguide shown in Fig. 26b.

In this structure, an advanced fabrication technique is required to form closely separated twin waveguides. Waveguide ridges are formed by using electron-cyclotron-resonance (ECR) etching with $SiCl_4$. Figure 30 is a set of views of the two adjacent waveguides in the coupling region. The waveguides and electrodes are clearly separated, in spite of the fact that they are only separated by 1 μm.

The operation characteristics of the fabricated directional-coupler-type switch are summarized in Fig. 31. 'Output A' and 'Output B' are indicated

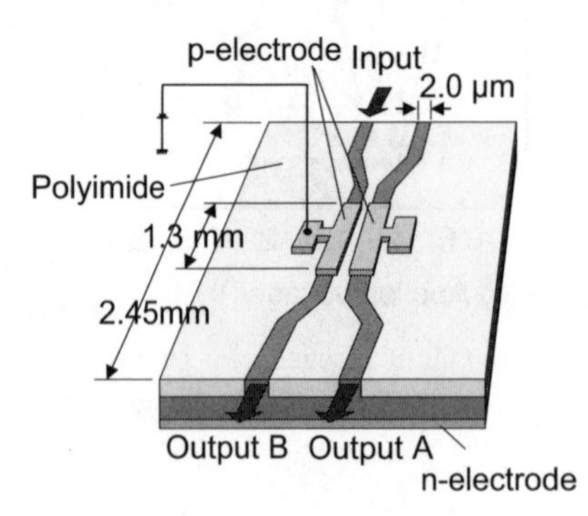

Fig. 29. Schematic illustration of a directional-coupler-type switch using polariton propagation

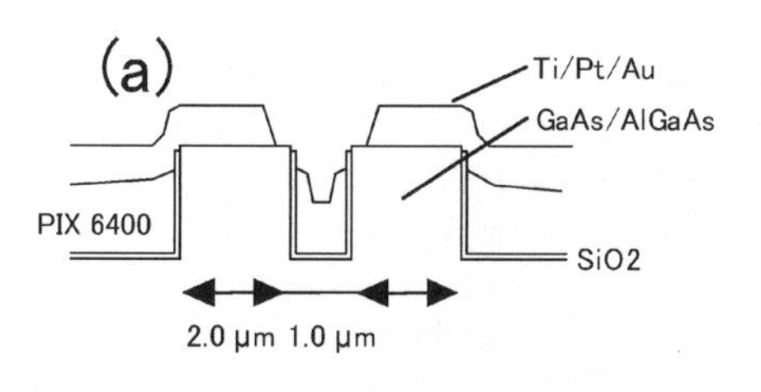

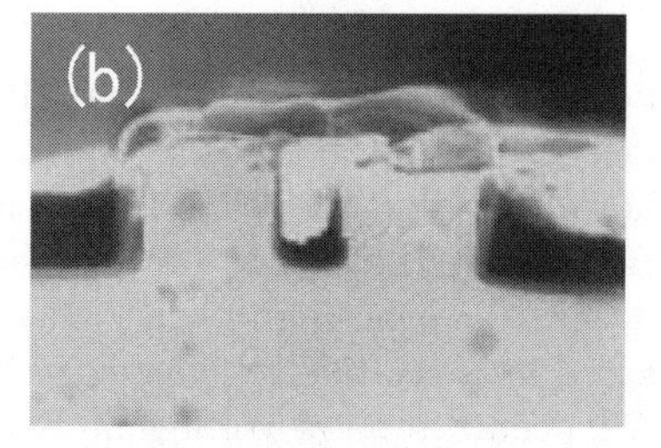

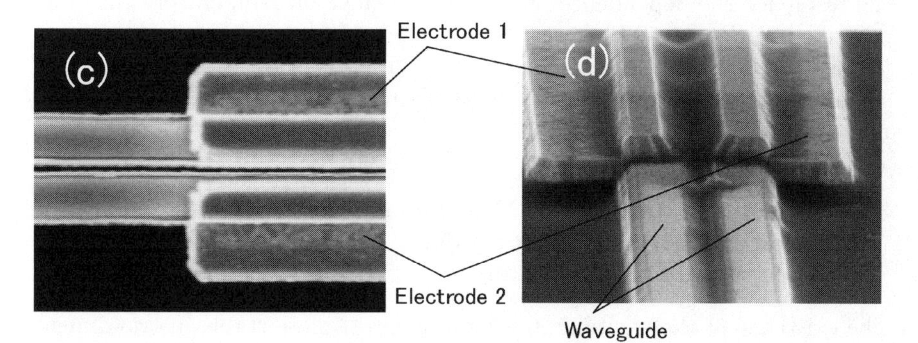

Fig. 30a-d. Fabricated coupling region of the directional-coupler-type switch: (**a**) a schematic cross-sectional view of the region, (**b**) a cross-sectional view of the region in the fabricated device, (**c**) a plan view of the coupling region, and (**d**) an overview of the coupling region. Figures (**b**)–(**d**) are SEM images

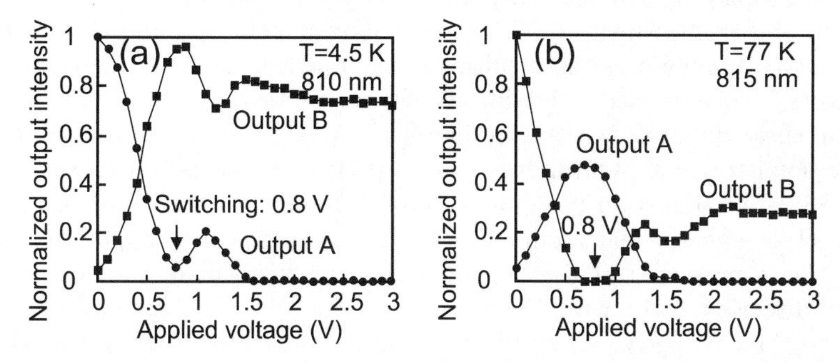

Fig. 31a,b. Voltage dependence of normalized output-light intensity for the directional-coupler-type switch at (**a**) 4.5 K and (**b**) 77 K. The amount of detuning wavelength was the same in both cases. The configuration of 'Output A' and 'Output B' is denoted in Fig. 29

in Fig. 29. The laser beam illuminates the input side of the waveguide that corresponds to 'Output B'. The voltage is applied to the illuminated waveguide. The characteristics measured at 4.5 K are shown in Fig. 31a. The 'Output A' intensity decreases rapidly with increasing voltage. On the other hand, the 'Output B' intensity increases rapidly. Switching occurs at 0.8 V, which corresponds to the signal switching voltage. Low-voltage switching had thus been demonstrated. Note that the intensity of 'Output B' reaches the same level as the initial 'Output A' intensity. This means that the output power is not affected by the transmission loss, which is caused by the exciton absorption. Thus, the extinction ratio in switching is a high 13 dB.

On the other hand, Fig. 31b shows the operation characteristics as measured at a relatively high temperature 77 K. As the figure shows, switching still occurs at a low voltage, in this case 0.8 V. The result again coincides with the result for the dependence of the phase-change on temperature that was described in Sect. 2.5. However, the 'Output A' intensity does not recover to the same level as the initial 'Output B' intensity. This is probably due to some defects in the device that will be eliminated after further study of the process technology involved.

3.3 Spatial Confinement of Electromagnetic Field by an Excitonic Polariton Effect: Theoretical Considerations

The existence of the excitonic polariton is much more stable in two-dimensional quantum structures (quantum wells) than in bulk materials, as was shown in the previous section. Therefore, excitonic polaritons in one-dimensional quantum structures (quantum wires) are expected to be even more stable, because the exciton oscillator strength is larger in this case than in quantum wells [56]. On the other hand, extremely thin wire structures such as semiconducting whiskers [57,58] and nanochannel array glasses [59] have been successfully fabricated, resulting in the realization of the quantum-wire structures. Consequently, studying the characteristics of the excitonic polaritons in these quantum wires is an important part of achieving a much more stable polariton state. Obtaining a more stable excitonic polariton will make the fabrication of further new optoelectronic devices possible. In this section, the electromagnetic field profile of the excitonic polariton in a quantum-wire structure is presented. This leads to strong confinement of the field in the lateral direction of the wire.

First, the refractive index of the excitonic polariton in a quantum-wire structure is calculated. The range of materials considered here consists solely of GaAs, because this material is widely used in quantum-confined systems. Assuming that the material is isotropic and has a permeability constant of 1, we can obtain the following dispersion relation for plane electromagnetic (light) waves:

$$\frac{c^2k^2}{\omega^2} \approx \varepsilon_0 + \frac{4\pi a_0 \omega_0^2}{\omega_0^2 - \omega^2 + (\hbar k^2 \omega_0 / m^*) - \mathrm{i}\omega\Gamma} \,. \tag{14}$$

In this equation, k is a wavevector, ω is the frequency, ε_0 is the background dielectric constant, a_0 is the oscillator strength, ω_0 is the resonant frequency,

m^* is the exciton mass, Γ is the damping factor, c is the velocity of light in vacuum, and $\hbar$ is Planck's constant ($\hbar = h/2\pi$). The transverse exciton frequency ω_t was approximated by the parabolic dependence on k, as

$$\hbar\omega_t \approx \hbar\omega_0 + \frac{1}{2}\frac{h^2k^2}{m^*} \,. \tag{15}$$

Therefore, we can obtain k^2 by solving (14). Since $n^2 = c^2k^2/\omega^2$, we can obtain two sets of refractive indices n_u and n_l, which correspond to the two polariton branches (upper-branch polariton and lower-branch polariton). The effective refractive index n can be obtained by using the boundary condition that the total polarization is 0 at the boundary of the system [60],

$$n = (n_u n_l + \varepsilon_0)/(n_u + n_l) \,. \tag{16}$$

Here, the polariton stability can be measured as the difference between the longitudinal frequency ω_l and the transverse exciton frequency ω_t, which is called the LT splitting energy, Δ.

$$\Delta = \hbar\omega_l - \hbar\omega_t = \left[\left(1 + \frac{4\pi a_0}{\varepsilon_0}\right)^{1/2} - 1\right]\hbar\omega_t \,. \tag{17}$$

The LT splitting Δ can be estimated from the exciton oscillator strength, because they are proportional to each other, as is indicated by (11). The relationship between the oscillator strength of the GaAs exciton and the confining dimensionality was numerically calculated by Matsuura and Kamizato [56]. They calculated the oscillator strengths for quantum-well and wire structures with infinite barrier potential by using the variational model. According to their calculation, the oscillator strength increases rapidly as the well and wire widths decrease. For example, the oscillator strength for a 5-nm GaAs quantum wire is 19 times larger than that for bulk GaAs, while that for a 10-nm GaAs quantum wire is 5.6 times larger. Therefore, we can estimate Δ to be 0.56 and 1.9 meV for 10- and 5-nm wide GaAs quantum wires, respectively, because the value of Δ for the bulk GaAs is about 0.1 meV.

The refractive index of a GaAs quantum wire as calculated by using (14) and (16) is shown in Fig. 32. The calculation used $\Delta = 2$ meV as a typical value for a GaAs quantum-wire structure. The damping factor Γ was assumed to be 0.01 meV because the several reported values are of the order of 0.01 meV [61–63]. The background dielectric constant ε_0 is 12.53 [64], the exciton mass m is $0.517m_0$ (m_0: electron mass in vacuum) [44], and the resonant frequency ω_0 is 1.622 eV (GaAs bandgap) [24]. As is shown in Fig. 32, the real part of the refractive index reaches its maximum at an energy that is slightly lower than that for the resonant frequency ω_0. The maximum refractive index is as large as 5.6, which is about 1.6 times larger than that of bulk GaAs. The minimum value, on the other hand, which is less than unity, is located at around $\hbar\omega_0 + \Delta$. Note, however, that the absolute value

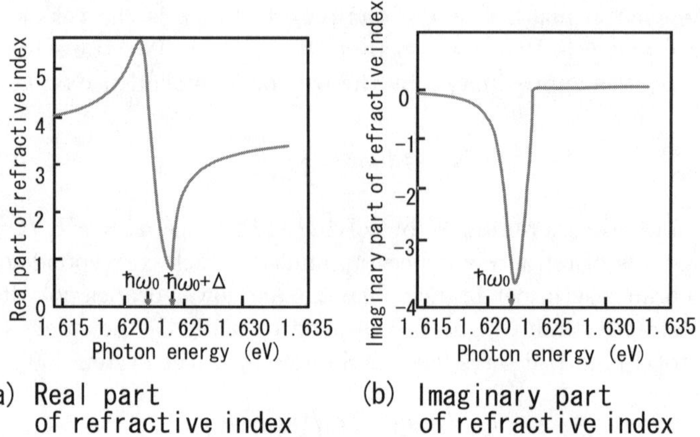

Fig. 32. Calculated refractive index for a GaAs quantum-wire structure; (**a**) real part of refractive index and (**b**) imaginary part of refractive index. The LT splitting Δ is 2 meV, and the damping factor Γ is 0.01 meV. The resonant frequency is ω_0

of the imaginary part of the refractive index increases rapidly as the energy approaches ω_0. The maximum value is thus located at an energy level slightly higher than ω_0.

The behavior of the real part of the refractive index of the GaAs quantum-wire structure is shown in Fig. 33. In the figure, the upper line shows the maximum value of the real part of the refractive index while the lower line shows the minimum value. The maximum value of the real part of the refractive index increases rapidly with the LT splitting, Δ. Such large refractive indices may be able to confine light to an extremely small area, because the

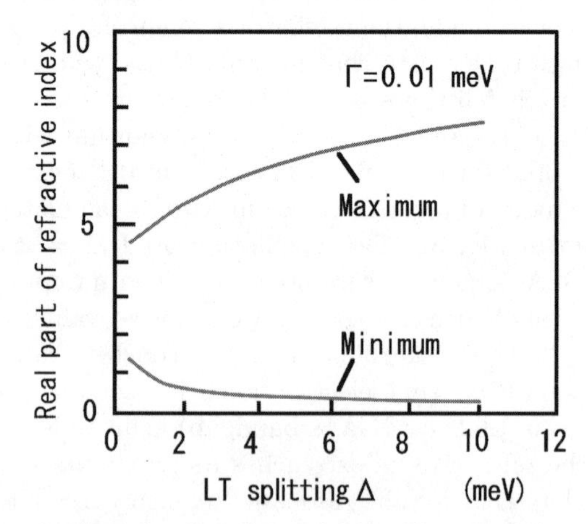

Fig. 33. Maximum and minimum values of the real part of the refractive index as a function of LT splitting. The damping factor Γ is 0.01 meV

wavelength of light in a material is inversely proportional to the refractive index. In order to examine the electromagnetic field of the polariton in the quantum wire, a three-layered slab waveguide is considered for the sake of simplicity. Furthermore, the profile of the electromagnetic field is calculated on the assumption of the continuum model, although the quantum wire is extremely narrow. This assumption has been found to be reasonable by Philpott [65] and Nozue et al. [66]. For example, Philpott [65] calculated the refractive index of a layer of point dipoles and obtained a form of the refractive index equivalent to that of the continuum model.

The structure of the three-layered slab waveguide is shown in Fig. 34. This structure is characterized by its refractive-index profile (n_0 core, n_1 cladding). When an electromagnetic wave of frequency ω travels in the form of $\exp[\mathrm{i}(\omega t - \beta z)]$ and the confinement is only in the x direction (t is time, β is the propagation constant in the z direction), the electric field is given by

$$\begin{cases} A(\mathrm{e}^{\kappa x} \pm \mathrm{e}^{-\kappa x}) & (|x| \leq L/2)\,, \\ B(\mathrm{e}^{-\sigma x}) & (|x| > L/2)\,, \end{cases} \tag{18}$$

where A, B, κ, and σ are complex, and κ and σ correspond to phase constants. In the first line of (18), + and − correspond to even and odd modes, respectively. The following eigenfunction is derived under the conventional boundary condition:

$$\frac{\mathrm{e}^{\kappa L/2} \mp \mathrm{e}^{-\kappa L/2}}{\mathrm{e}^{\kappa L/2} \pm \mathrm{e}^{-\kappa L/2}} = -\frac{\sigma}{\kappa} N\,, \tag{19}$$

where N is defined as $N = 1$ for the TE mode, and $N = (n_0/n_1)^2$ for the TM mode. The upper signs in (19) correspond to the even mode and the lower signs correspond to the odd mode. In the TE mode, the polarization is in a plane along the y axis, while in the TM mode, the polarization is perpendicular to the plane. Furthermore,

$$(n_0 k_0)^2 + \kappa^2 = (n_1 k_0)^2 + \sigma^2 = \beta^2 (\equiv k_0 n_{\mathrm{eff}})\,, \tag{20}$$

where $k_0 = 2\pi/\lambda$ (λ is the wavelength in a vacuum) and n_{eff} is the effective refractive index. The electric field of each mode can be obtained by deriving β from (19) and (20).

The following structures of the waveguides were used in the calculation.

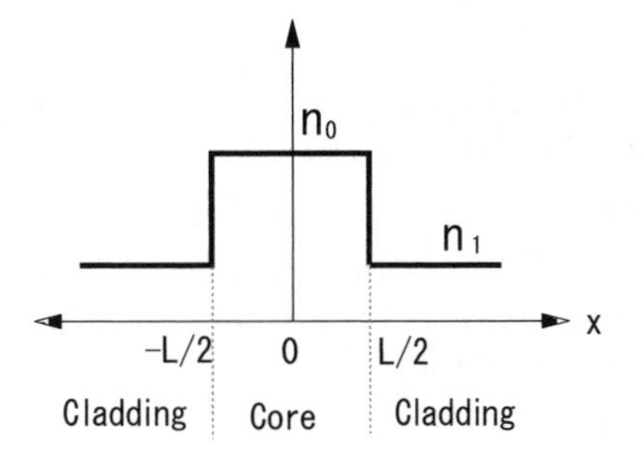

Fig. 34. Refractive-index profile of the three-layered slab waveguide

1. The core is composed of a GaAs quantum-wire structure, and the cladding is conventional bulk GaAs.
2. The core is composed of a GaAs quantum-wire structure, and the cladding is air.

The second case corresponds to a thin wire crystal of GaAs [57,58]. In the calculation, the core width is varied to obtain the corresponding dependence of the electromagnetic field profile on the core width. Therefore, it is assumed that multiple quantum wires provide a model for the core structure when its width is increased. In this case, the multiple-wire structure can be treated as an effective-average medium. Of course, the wire spacing should be determined so as to avoid the interaction of carriers between wires. Then, the refractive index of the core is selected as the maximum value of the real part of the refractive index of the excitonic polariton, which appears at an energy slightly lower than the resonant frequency ω_0. Figure 35 shows the power (magnitude of the poynting vector) profile for case 1, i.e., where the cladding is bulk GaAs. The core width L is 20 nm, and the LT splitting is fixed at 2 meV. The mode is the TE fundamental mode. The figure shows that the full width at half maximum (FWHM) of the power profile is small, at only 67 nm. Figure 36 shows the relationship between the FWHM of the power profile and LT splitting Δ. The core width L is varied from 5 to 30 nm. The FWHM of the power profile becomes smaller as Δ and L are increased.

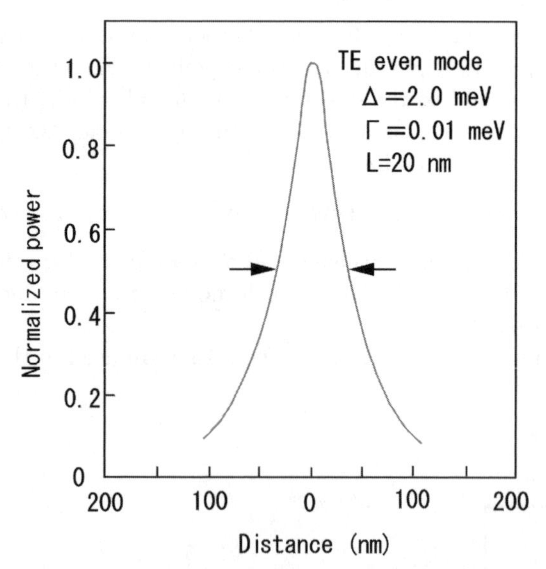

Fig. 35. Power profile in the structure where the core has a GaAs quantum-wire structure and the cladding is of conventional bulk GaAs. The LT splitting Δ is 2 meV, the damping factor Γ is 0.01 meV, and the core width L is 20 nm. Calculation is for the TE even mode

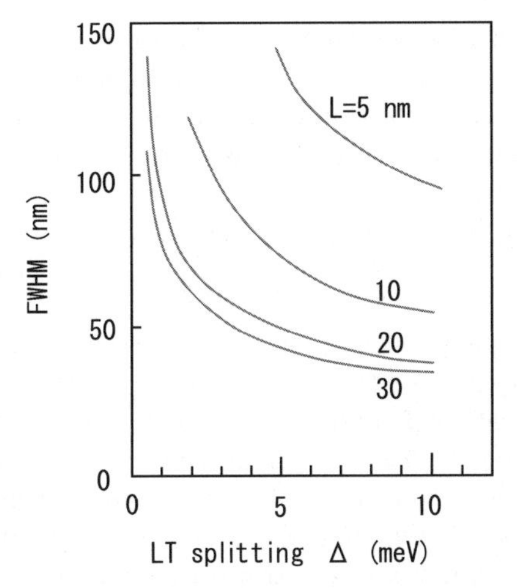

Fig. 36. Full width at half maximum (FWHM) of the power profile as a function of LT splitting in the structure where the core has a GaAs quantum-wire structure and the cladding is for conventional bulk GaAs. The parameter is the core width

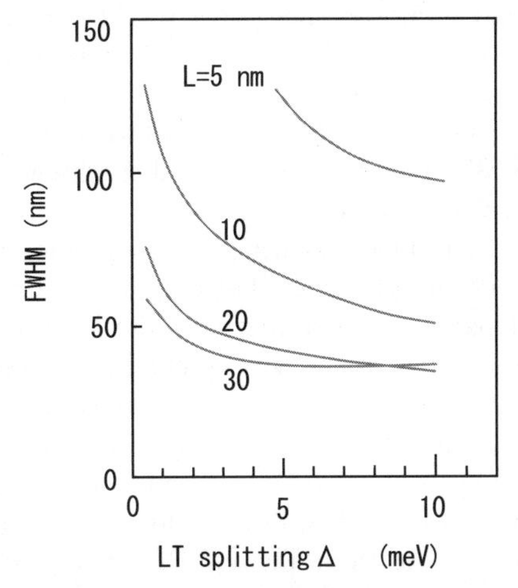

Fig. 37. Full width at half maximum (FWHM) of the power profile as a function of LT splitting. The structure used in the calculation has a core of GaAs quantum wire and cladding of air. The parameter is the core width

However, the decrease in the FWHM of the power profile reaches saturation when the core width reaches 20 nm.

On the other hand, Fig. 37 shows the behavior of the power profile in case 2, where the cladding is air with a refractive index of 1. The FWHM of the power profile is much smaller than in case 1. This is due to the large difference between refractive indices of the core and cladding.

Therefore, the light power related to the excitonic polariton is strongly confined by the increase in refractive index. Note, however, that the transmission loss of the excitonic polariton propagation is also increased because the imaginary part of the refractive index is high at the energy position where the real part of the refractive index is at its maximum. The transmission loss α can be approximated by a linear combination of the loss in each layer;

$$\alpha = \alpha_{\mathrm{I}} f_{\mathrm{I}} + \alpha_{\mathrm{II}} f_{\mathrm{II}} , \tag{21}$$

where α_{I} and α_{II} are the losses in the quantum-wire core and cladding, respectively, as directly derived from the imaginary part of the refractive index of each layer. The values f_{I} and f_{II} are the fraction of the light power in the quantum-wire core to the total light power and the equivalent fraction for the cladding, respectively. The calculated transmission loss for the light-power profile described in Fig. 35 ($\Delta = 2$ meV, $\Gamma = 0.01$ meV, $L = 20$ nm) is large, at 1×10^7 m^{-1}. However, the transmission loss can be reduced by detuning the energy from the resonant-energy position. If the energy is detuned by 3.5 meV from the energy position where the real part of the refractive index is at its maximum, the transmission loss is reduced to only 3×10^5 m^{-1}. Furthermore, it should be noted that the FWHM of the power profile does not greatly increase, even though the transmission loss is significantly reduced. When the transmission loss is 3×10^5 m^{-1}, the FWHM of the power profile is still 180 nm, which is only 2.7 times larger than the FWHM of the power profile in Fig. 35. This is due to the slow change in the real part of the refractive index relative to the change in the imaginary part. The transmission loss can also be reduced by using a waveguide structure with a core of conventional bulk GaAs and with cladding layers that are quantum-wire structures. The real part of the refractive index of the cladding quantum wire should be less than that of the core to confine the light power. We used the minimum value of the real part of the refractive index. The value is 0.6, which is smaller than unity. In this case, the light power is confined in the lossless core region. Therefore, a relatively small loss of 2.5×10^6 m^{-1} can be obtained in a waveguide with a 40-nm wide core, even when the energy is exactly at the resonant position. The FWHM of the power profile remains small (65 nm). Furthermore, if the energy is detuned by 4 meV from the energy position that gives the minimum refractive index, the transmission loss is reduced to only 1×10^4 m^{-1}. On the other hand, the FWHM of the power profile is still small, at only 170 nm. Therefore, it is possible to reduce the transmission loss while keeping the FWHM of the power profile small.

In summary, the refractive index related to the excitonic polariton in a GaAs quantum wire has been discussed on the basis of theory. The maximum value of the real part of the refractive index was found to be as large as 5.6 when the width of the GaAs wire is 5 nm. The electromagnetic field of the excitonic polariton transmitted in such a quantum-wire structure is therefore strongly confined because the wavelength in the structure is reduced by the increase in refractive index. The estimated FWHM of the light power is typically small, at only 67 nm, although the transmission loss is fairly high. However, the transmission loss can be reduced by detuning the wavelength from the resonant energy, while keeping the FWHM of the power profile small. These results suggest the possibility of making extremely small optical devices.

3.4 Nanometer-Scale Switches

As shown in Sect. 3.3, the profile of the electromagnetic field of the excitonic polariton is spatially squeezed in a quantum-wire structure (two-dimensional confinement). This result is based on the large refractive index, which is caused by the large quantum-confinement effect. Such squeezing of the field is particularly important for device applications, because it raises the possibility of ultrasmall optoelectronic devices, i.e., of devices with sizes of the order of a micrometer [30].

In this section, we discuss the propagation of excitonic polaritons in quantum-wire waveguides in adjacent-parallel configurations. Such parallel waveguides are particularly important for obtaining such active devices as directional coupler switches. We used simulation based on the finite-element method to study the propagation behavior of the excitonic polaritons.

The simulation is based on the continuum model. Therefore, we derive the refractive index of the waveguide from the dispersion relation of the excitonic polariton. It is then assumed that the propagation of the excitonic polariton is equivalent to the propagation of an electromagnetic wave in a waveguide with a refractive index derived from the above procedure. As was discussed in Sect. 3.3, this continuum model has been demonstrated to be reasonable, although the width of a quantum-wire waveguide is extremely small [65]. The details of the deviation of the refractive index were given in Sect. 3.3 [39].

The procedure we used for simulation based on the finite-element method is as follows. We consider an electromagnetic wave with a single frequency ω and time dependence of the form $\exp(\mathrm{i}\omega t)$. Furthermore, the variation in refractive index over distances of the order of one wavelength is negligibly small within the regions surrounded by fixed structural boundaries. Note, however, that discontinuity of the refractive index is allowed when the boundaries are crossed. In other words, $\nabla \ln \varepsilon = \delta\varepsilon/\varepsilon \approx 0$ for the dielectric constant ε, and the following Helmholtz equation is derived from Maxwell's equation;

$$\nabla^2 E(x,y,z) + k_0^2 n^2(x,y,z) E(x,y,z) = 0 . \tag{22}$$

Here, $E(x,y,z)$ is the one scalar component of the electric field, k_0 is the free-space wavenumber, and $n(x,y,z)$ is the refractive index. $E(x,y,z)$ and $n(x,y,z)$ are complex numbers that represent phase and loss. The propagation direction is the z axis. We introduce the "slowly varying envelope approximation":

$$E(x,y,z) = \Phi(x,y,z) \exp(-\mathrm{i}k_0 n_0 z) , \tag{23}$$

where n_0 is a constant refractive index. Furthermore, we neglect the reflection that is generated by the existence of the incident wave, so we introduce the following condition:

$$\left| \frac{\partial^2 \Phi(x,y,z)}{\partial z^2} \right| \ll \left| 2k_0 n_0 \frac{\partial \Phi(x,y,z)}{\partial z} \right| . \tag{24}$$

Then, we can obtain

$$2\mathrm{i}k_0 n_0 \frac{\partial \Phi(x,y,z)}{\partial z} = \nabla_t^2 \Phi(x,y,z) + k_0^2 (n^2(x,y,z) - n_0^2) \Phi(x,y,z) , \tag{25}$$

where $\nabla_t = \nabla - t(0,0,\partial/\partial z)$. As a result, we obtain the relation between $\Phi(x,y,z)$ and $\Phi(x,y,z+\delta z)$ in the following way:

$$\begin{aligned} \Phi(x,y,z+\delta z) &= \Phi(x,y,z) + \delta z \frac{\partial \Phi(x,y,z)}{\partial z} + O(\delta z^2) \\ &= \left[1 - \mathrm{i} \frac{\delta z}{2k_0 n_0} \left\{ \nabla_t^2 + k_0^2 (n^2(x,y,z) - n_0^2) \right\} \right] \Phi(x,y,z) \\ &\quad + O(\delta z^2) . \end{aligned} \tag{26}$$

Therefore, we can successively calculate the profile of the electric field $E(x,y,z+\delta z)$ in the direction of propagation $z+\delta z$ from the profile $E(x,y,z)$ at z.

We used DEQSOL E2 (differential equation solver language extended version 2) for the numerical calculations. We started by evaluating the accuracy of the simulation by comparing the analytical calculation. The comparison was made by using a single three-layered slab waveguide. In this case, the refractive index only varies along the x axis. The core consists of a GaAs quantum-wire structure and the cladding is conventional bulk GaAs, i.e., the structure is the same as that described in Sect. 3.3 [39]. The refractive index of the core is assumed to be $5.56 - 2.24\mathrm{i}$, that corresponds to the refractive index for the resonant energy (1.622 eV). The LT splitting is fixed at 2 meV [39]. The refractive index of the bulk GaAs used for the cladding is fixed at 3.55 [39]. The core width is 10 nm, and the waveguide length is 1 μm. We then considered the TE mode that corresponds to $E_z = E_x = 0$, and found that an accurate profile of the electric field is obtained when the mesh number of the core and the step number of the z axis are both more than 100.

Next, we simulated the propagation of the excitonic polariton along the adjacent-parallel waveguides. Figure 38 shows the waveguide structure and the corresponding refractive index profile. In this case, again, we used the slab waveguide and the TE-mode configuration. The waveguide width is 40 nm and the distance between waveguides is 160 nm. In the calculation, the incident photon energy was fixed to the resonant region of the polariton (1.622 eV). Therefore, the refractive index of the core was $5.56 - 2.24\mathrm{i}$ and that of the cladding was 3.55 [39]. The power profile of the incident light was

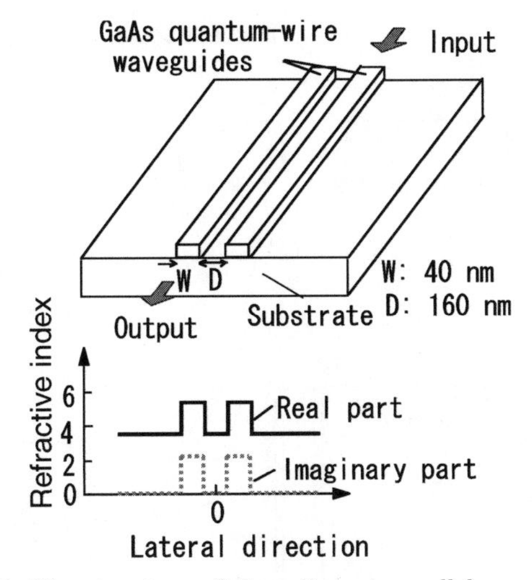

Fig. 38. The structure of the adjacent-parallel waveguides

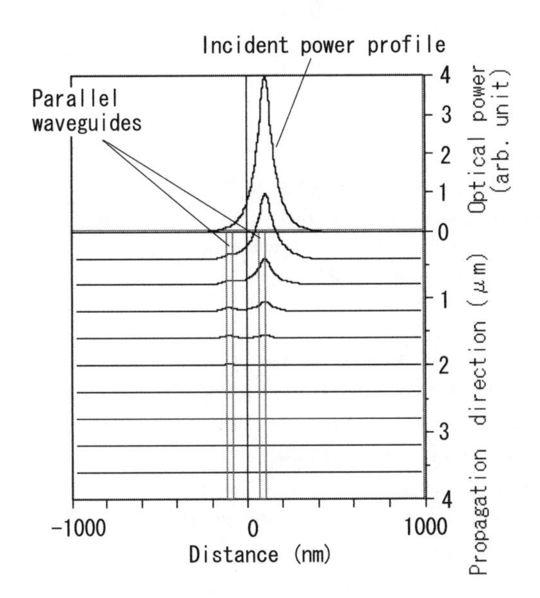

Fig. 39. Power profile for the excitonic polariton propagating in the adjacent parallel GaAs quantum-wire waveguides. The incident photon energy is fixed at the resonance energy of 1.622 eV. The waveguide is 40 nm wide and the distance between the waveguides is 160 nm

assumed to be identical to that of the fundamental mode of the three-layered slab waveguide, and the peak position of the power profile was set at the center of the right-hand waveguide.

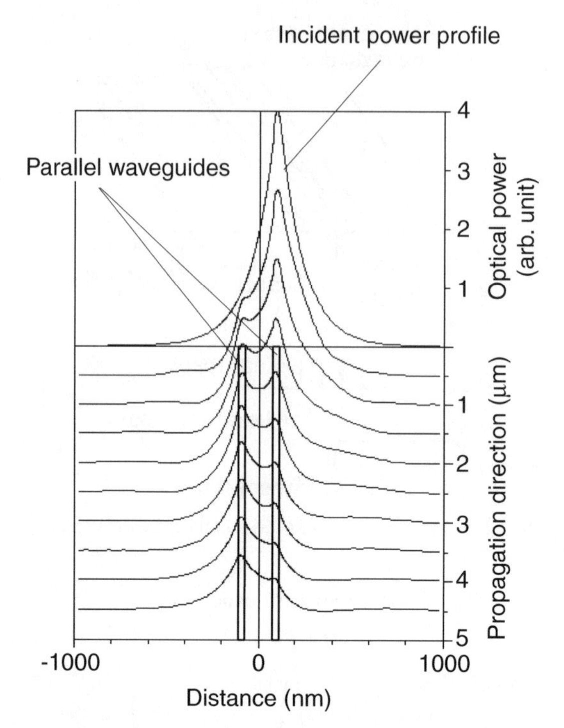

Fig. 40. Power profile for the excitonic polariton propagating in the adjacent-parallel GaAs quantum-wire waveguides. The incident photon energy is fixed at 1.614 eV, which is detuned by 8 meV from the resonance energy. The other parameters are the same as those in Fig. 39

The calculated power profile of the propagating excitonic polariton is shown in Fig. 39. As shown, the second peak grows at the center of the left-side waveguide as the field propagates along the waveguides. The power is transferred from the right-hand to the left-hand waveguide across the small distance of 2.5 μm. However, the peak power falls to only 1% of the peak level at the input end. This is because severe attenuation of the power occurs due to the resonance energy region.

We can avoid this severe attenuation of the propagating excitonic polariton by detuning the incident photon energy. Figure 40 shows the result for the photon energy of 1.614 eV, which is detuned by 8 meV from the resonance energy. In this case, we used $3.96 - 0.077i$ as the refractive index of the core. Although the maximum rate of the power conversion is across the slightly greater distance of 4 μm, there is much less reduction in peak power than in the resonance case. The peak power only falls to 27% of its level at the input end.

In summary, we have demonstrated the use of a pair of adjacent-parallel GaAs quantum-wire waveguides to achieve a directional coupler that has a very short coupling length. A typical coupling length is 4 μm. Since the average coupling length of the conventional optical directional coupler is of the order of 1 mm [67], the use of the excitonic polariton offers a coupling length three orders of magnitude shorter than the conventional case. Such short coupling lengths will make it possible to fabricate extremely small directional-coupler-type switches.

4 Summary and Future Prospects

We have reviewed fundamental aspects of excitonic polaritons propagating in such quantum-confined structures as quantum-well waveguides. We also discussed the possible applications of excitonic polaritons in optoelectronic devices. We commence the summary by reviewing the results we obtained with regard to fundamental aspects of excitonic polaritons propagating in quantum-well waveguides.

1. The propagation of the excitonic polaritons in GaAs quantum-well waveguides was demonstrated by using time-of-flight measurements. The group velocity of the transmitted light pulse associated with the heavy-hole exciton was 3.7×10^4 m/s at liquid-helium temperature, while that associated with the light-hole exciton was 4.2×10^4 m/s. These large reductions in group velocity are direct evidence of the existence of the excitonic polariton in a GaAs quantum-well waveguide. This polariton effect is significantly enhanced in the configuration where the polarization of the incident light is perpendicular to the quantum well. The group velocity of the polaritons was found to increase and approach that of light with increasing incident laser power, since this led to increased polariton scattering.
2. Propagation of excitonic polaritons in a quantum-well waveguide was theoretically investigated by first-principles calculation of the spatial-dispersion relations of an excitonic polariton. This calculation was based on an additional-boundary-condition-free theory and guided-mode characteristics. The results were in reasonable agreement with time-of-flight measurements, confirming the existence of the quantum-well excitonic polariton.
3. The dependence of the propagation of excitonic polaritons in a quantum-well waveguide on electric field was investigated by interferometric spectroscopy. A large interference-fringe shift was observed, showing that the electric field was causing a large change in refractive index. Two-dimensional enhancement of the photon–exciton interaction caused an increase in the LT splitting energy of the quantum-well exciton to five times the value for a three-dimensional bulk exciton.
4. The temperature dependence of the polariton phase-change under an electric field was investigated by interferometry. The rate of phase-change is considerably larger at lower than at higher temperatures. The relatively

large phase-change remains in place at temperatures up to 120 K. This critical temperature is much higher than we had expected, and the physical mechanism responsible for this is not clear at present. This high critical temperature indicates the possibility of polariton device operation at relatively high temperatures.

5. We evaluated whether waveguide structures with a grating are valid for obtaining strong coupling between excitons and photons. The energy of interaction in the resulting structures is 3.1 meV, which is consistent with the results of calculation with a model based on the dispersion relation of polariton. This value is seven times larger than the magnitude of the LT splitting, and we were able to observe, at relatively high temperatures, double-peak structures in PL spectra that are due to coherent exciton–photon interaction.

We now summarize our results with regard to device applications of the excitonic polaritons propagating in quantum-well/wire waveguides.

1. The operation characteristics, at low temperatures, of Mach–Zehnder-type modulators and directional-coupler-type switches were investigated. The principles of operation applied in these devices are based on phase-changes under an electric field. The waveguides used for the devices were composed of GaAs/GaAlAs quantum wells. Low operation voltages, i.e., less that 1 V, were obtained for devices of both types at least up to 77 K. This characteristic coincides with the temperature dependence of the phase-change observed directly by an interferometric configuration using a straight waveguide.
2. In the case of one-dimensional quantum-confined systems, i.e., quantum wires, the refractive index related to the excitonic polariton in a GaAs quantum wire was calculated numerically. The maximum value of the real part of the refractive index was large, at 5.6, when the width of the GaAs wire was 5 nm. This large refractive index was a result of the large binding energy of the exciton confined in a one-dimensional structure. The electromagnetic mode profile of the excitonic polariton transmitted in such a quantum-wire structure is spatially squeezed by the reduction of the wavelength in the structure that is, in turn, produced by the increase in the refractive index. The estimated full width at half maximum (FWHM) of the light power related to the excitonic polariton was small, with 67 nm as a typical value.
3. We have demonstrated a very short coupling length of the directional coupler consisting of adjacent-parallel GaAs quantum-wire waveguides. A typical coupling length was 4 μm. The average coupling length for a conventional optical directional coupler is of the order of 1 mm, so applying the excitonic polariton promises a three-order-of-magnitude decrease. Such short coupling lengths will make it possible to fabricate extremely small directional-coupler-type switches.

As shown above, excitonic polaritons in quantum-confined systems such as quantum wells and wires can be transmitted as guided waves along the waveguide structure. Furthermore, they exhibit properties of both charged electrons (holes) and light waves. Excitonic polaritons thus have excellent characteristics such as (1) excellent coherence as a wave state, (2) a large phase-change under an electric field, and (3) the capacity for transmission along an extremely thin waveguide. By analogy with such conventional optical devices as optical waveguides and optical switches, these characteristics lead to the realization of optoelectronic devices with ultralow operation voltages and/or ultrasmall sizes.

However, the existence of the excitonic polariton was thought to be limited to relatively low temperatures. Although we found that the induction of the phase-change by an electric field in the GaAs/AlGaAs quantum-well waveguide remains strong at temperatures up to 120 K, this is still within the low-temperature region. Low-temperature devices are, of course, useful in some fields of application. However, we expect to be able to obtain room-temperature operation, which will open up such major fields of application as optical network systems.

As described here, we have described our demonstration of a new basis for devices that handle the excitonic polariton. The next step should be the actual application of polariton-based devices. These devices should operate at higher temperatures, including room temperature. In order to increase the operation temperature, much higher stability is required for polaritons. The way to achieve this increase in stability is to increase the LT (longitudinal-transverse) mode splitting energy of the polariton. Fortunately, many of new materials with large values for LT-splitting energy have recently been found. For example, PbI_4-based layered perovskite-type semiconductors have a large polariton mode splitting energy of 100 meV, which means that the polariton is stable, even at room temperature [68]. Another way to increase the LT-splitting energy is to introduce high-quality quantum-wire structures. Various nanofabrication techniques are now under investigation. The development of such new materials and fabrication techniques will lead to the realization of polariton-based devices that operate at much higher temperatures, including room temperature.

Acknowledgments

The authors wish to thank K. Ogawa, M. Shirai, T. Sato, H. Nakamura, M. Kawata, S. Nishimura, J. Shigeta, T. Mishima, K. Hiruma, T. Shimada, K. Oimatsu, and T. Iida for their important contributions to this work. A part of this work was carried out by the Quantum Functional Devices Project under the management of FED (the R&D Association for Future Electron Devices) as a part of the METI (Ministry of Economy, Trade and Industry) R&D of Industrial Science and Technology Frontiers program supported by

NEDO (the New Energy and Industrial Technology Development Organization).

References

1. K. Cho: 'Excitons'. In: *Topics in Current Physics* **14** (Springer, Berlin 1979)
2. E.I. Rashba, M.D. Sturge: In: *Excitons, Modern Problems in Condensed Matter Sciences* **2**, ed. by V.M. Agranovich, A.A. Maradudin (North Holland, Amsterdam 1982)
3. M. Ueta, H. Kanzaki, K. Kobayashi, Y. Toyozawa, E. Hanamura: In: *Excitonic Processes in Solids, Springer Series in Solid-State Science* **60**, ed. by M. Cardona, P. Fulde, K. von Klitzing, H.-J. Queisser (Springer, Berlin 1986)
4. S.I. Pekar: Zh. Eksp. Teor. Fiz. **6**, 1022 (1958) [Sov. Phys. JETP **6**, 785 (1958)]
5. J.J. Hopfield: Phys. Rev. **112**, 1555 (1958)
6. J.J. Hopfield, D.G. Thomas: Phys. Rev. **132**, 563 (1963)
7. T. Mita, K. Sotome, M. Ueta: J. Phys. Soc. Jpn. **48**, 496 (1980)
8. G. Winterling, E.S. Koteles: Solid State Commun. **23**, 95 (1977)
9. R.G. Ulbrich, C. Weisbuch: Phys. Rev. Lett. **38**, 865 (1977
10. Y. Masumoto, Y. Unuma, Y. Tanaka, S. Shionoya: J. Phys. Soc. Jpn. **47**, 1844 (1979)
11. R.G. Ulbrich, G.W. Fehrenbach: Phys. Rev. Lett. **43**, 963 (1979)
12. Y. Nozue: J. Phys. Soc. Jpn. **51**, 1840 (1982)
13. R.L. Greene, K.K. Bajaj: Solid State Commun. **45**, 831 (1983)
14. T. Ishibashi, S. Tarucha, H. Okamoto: In: *IOP Conference Proceedings*, No. 63 (Institute of Physics, Bristol, London 1982) pp. 587
15. T.H. Wood, C.A. Burrus, D.A.B. Miller, D.S. Chemla, T.C. Damen, A.C. Gossard, W. Wiegmann: Appl. Phys. Lett. **44**, 16 (1984)
16. D.A.B. Miller, D.S. Chemla, T.C. Damen, A.C. Gossard, W. Wiegmann, T.H. Wood, C.A. Burrus: Phys. Rev. Lett. **53**, 2173 (1984)
17. A. Mysyrowicz, D. Hulin, A. Antonetti, A. Migus, W.T. Masselink, H. Morkoc: Phys. Rev. Lett. **56**, 2748 (1986)
18. A. von Lehmen, D.S. Chemla, J.E. Zucker, J.P. Heritage: Opt. Lett. **11**, 609 (1986)
19. D.S. Chemla, D.A.B. Miller: J. Opt. Soc. Am. B **2**, 1155 (1985)
20. M. Kohl, D. Heitmann, P. Grambow, K. Ploog: Phys. Rev. B **37**, 10927 (1988)
21. E.L. Ivchenko, V.P. Kochereshko, P.S. Kop'ev, V.A. Kosobukin, I.N. Uraltsev, D.R. Yakovlev: Solid State Commun. **70**, 529 (1989)
22. M. Nakayama: Solid State Commun. **55**, 1053 (1985); M. Nakayama, M. Matsuura: Surf. Sci. **170**, 641 (1986)
23. K. Ogawa, T. Katsuyama, H. Nakamura: Appl. Phys. Lett. **53**, 1077 (1988)
24. K. Ogawa, T. Katsuyama, H. Nakamura: Phys. Rev. Lett. **64**, 796 (1990)
25. K. Oimatsu, T. Iida, H. Nishimura, K. Ogawa, T. Katsuyama: J. Lumin. **48&49**, 713 (1991)
26. K. Cho: J. Phys. Soc. Jpn. **55**, 4113 (1986)
27. T. Katsuyama, K. Ogawa: In: *Proceedings of the 3rd International Symposium on Foundations of Quantum Mechanics* (Tokyo 1989) pp. 315
28. T. Katsuyama, K. Ogawa: Semicond. Sci. Technol. **5**, 446 (1990)

29. M. Shirai, K. Hosomi, K. Hiruma, J. Shigeta, T. Katsuyama: Nonlinear Opt. **18**, 363 (1997)
30. T. Katsuyama, K. Ogawa: J. Appl. Phys. (Appl. Phys. Rev.) **75**, 7607 (1994)
31. K. Ogawa, T. Katsuyama: J. Lumin. **53**, 391 (1992)
32. K. Ogawa, T. Katsuyama, M. Kawata: Phys. Rev. B **46**, 13289 (1992)
33. T. Katsuyama, K. Ogawa: In: *Proceedings of the 1st International Workshop on Quantum Functional Devices (QFD '97)* (Nasu Heights 1992) pp. 82
34. K. Hosomi, M. Shirai, T. Katsuyama: In: *Part of the Conference on Photonics Technology into the 21th Century, Semiconductors, Microstructures, and Nanostructures, Singapore* **3899** (SPIE, 1999) pp. 176
35. T. Katsuyama, K. Ogawa: In: *Extended Abstracts of the 22nd Conference on Solid State Devices and Materials* (Sendai 1990) pp. 63-3
36. T. Katsuyama: FED Journal **6**, Suppl. 2, 13 (1995)
37. K. Hosomi, M. Shirai, J. Shigeta, T. Mishima, T. Katsuyama: IEICE Trans. Electron. **E82-C**, 1509 (1999)
38. T. Katsuyama, K. Hosomi: Microelectron. Eng. **63**, 23 (2002)
39. T. Katsuyama, S. Nishimura, K. Ogawa, T. Sato: Semicond. Sci. Technol. **8**, 1226 (1993)
40. T. Katsuyama, T. Sato, Y. Yamamoto, N. Sagawa: Superlattices Microstruct. **20**, 59 (1996)
41. J.S. Weiner, D.S. Chemla. D.A.B. Miller, H.A. Haus, A.C. Gossard, W. Wiegmann, C.A. Burrus: Appl. Phys. Lett. **47**, 664 (1985)
42. J. Hegarty: Phys. Rev. B **25**, 4324 (1982)
43. D. Marcuse: Bell Syst. Tech. J. **50**, 1791 (1971)
44. C. Weisbuch, R.G. Ulbrich: In: *Light Scattering in Solids 111, Topics in Applied Phys.* **51**, ed. by M. Cardona, G. Guntherodt (Springer, Berlin 1982) pp. 218
45. D. Marcuse: *Theory of Dielectric Optical Waveguides* (Academic, New York 1974)
46. M. Matsuura, Y. Shinozuka: J. Phys. Soc. Jpn. **53**, 3138 (1984)
47. J.E. Zucker, T.L. Hendrickson, C.A. Burrus: Appl. Phys. Lett. **52**, 945 (1988)
48. M. Shinada, S. Sugano: J. Phys. Soc. Jpn. **21**, 1936 (1966)
49. Y. Shinozuka, M. Matsuura: Phys. Rev. B **28**, 4878 (1983)
50. J.E. Zucker, T.L. Hendrickson: Appl. Phys. Lett. **52**, 945 (1988)
51. C. Weisbuch, M. Nishioka, A. Ishikawa, Y. Arakawa: Phys. Rev. Lett. **69**, 3314 (1992)
52. R. Houdre, C. Weisbuch, R.P. Stanley, U. Oesterle, P. Pellandini, M. Ilegems: Phys. Rev. Lett. **73**, 2043 (1994)
53. J.-K. Rhee, D.S. Citrin, T.B. Norris, Y. Arakawa, M. Nishioka: Solid State Commun. **97**, 941 (1996)
54. K. Hosomi, M. Shirai, T. Katsuyama: In: *4th International Workshop on Quantum Functional Devoces (QFD2000)* (Kanazawa 2000) pp. 117
55. K. Hosomi, M. Shirai, T. Katsuyama: In: *Conference on Optoelectronic and Microelectronic Materials and Devices* (La Trobe University 2000) pp. 113
56. M. Matsuura, T. Kamizato: Surf. Sci. **174**, 183 (1986)
57. K. Hiruma, M. Yazawa, T. Katsuyama, K. Ogawa, K. Haraguchi, M. Koguchi, H. Kakibayashi: J. Appl. Phys. **77**, 447 (1995)
58. T. Katsuyama, K. Hiruma, K. Ogawa, K. Haraguchi, M. Yazawa: Jpn. J. Appl. Phys. **34-1**, 224 (1995)
59. R.J. Tonucci, B.L. Justus, A.J. Campillo, C.E. Ford: Science **258**, 783 (1992)

60. S.I. Pekar: Sov. Phys. Solid State, **4**, 953 (1962)
61. L. Schultheis, I. Balslev: Phys. Rev. B **28**, 2292 (1983)
62. L. Schultheis, J. Lagois: Phys. Rev. B **29**, 6784 (1984)
63. D.D. Sell, S.E. Stokowski, R. Dingle, J.V. DiLorenzo: Phys. Rev. B **7**, 4568 (1973)
64. K.G. Hambleton, C. Hilsum, B.R. Holeman: Proc. Phys. Soc. **77**, 1147 (1961)
65. M.R. Philpott: J. Chem. Phys. **61**, 5306 (1974)
66. Y. Nozue, M. Kawaharada, T. Goto: J. Phys. Soc. Jpn. **56**, 2570 (1987)
67. H. Inoue, H. Nakamura, S. Sakano, K. Morosawa, T. Katsuyama, H. Matsumura: Optoelectronics **1**, 137 (1986)
68. T. Fujita, Y. Sato, T. Kuitani, T. Ishihara: Phys. Rev. B **57**, 12428 (1998)

Nano-Optical Imaging and Spectroscopy of Single Semiconductor Quantum Constituents

T. Saiki

1 Introduction

The optical control of single electronic quantum states is the most fundamental and critical technique needed for the functioning of nano-optical devices and for the implementation of quantum-information processing. Semiconductor quantum dots (QDs), where electrons are confined in a nanoscale volume, are one of the promising candidates for a prototype of such quantum systems due to their atom-like density of states, long duration of coherence [1,2] (narrow transition linewidth [3–5]), and strong interaction between confined carriers. So far, as a result of these specific properties in QDs, optical manipulations of single quantum states such as qubit rotations [6,7], optically induced entanglement [8], and single-photon turnstile [9] are realized in well-characterized QD systems including interface QDs formed in a narrow quantum well and self-assembled QDs grown in Stranski–Krastanow (S-K) mode.

Conventional far-field optical techniques are unsatisfactory for exploring and addressing these individual nanoscale systems in terms of spatial resolution, which is limited to half the wavelength of the light used due to the diffraction limit. Probing systems with fixed apertures or with mesa structures is a useful solution for isolating single-quantum constituents, but the imaging ability is sacrificed. Near-field scanning optical microscopy (NSOM), where single systems are observed through a small aperture at the end of a scanning probe, is a more powerful tool for locally accessing individual QDs and obtaining spatial information [10–12]. By achieving spatial resolution higher than 50 nm with reasonable sensitivity for spectroscopic measurements, we are able to explore internal structures of QDs; real-space mapping out of wavefunctions of electrons and holes (excitons) confined in the QDs.

The combination of femtosecond spectroscopy with NSOM offers new perspectives for the direct investigation of carrier dynamics [13,14] and the local manipulation of electronic quantum states [15,16] on the nanometer length scale. Time-resolved optical spectroscopy also provides a wealth of information on dynamic processes like the phase and energy relaxation of carriers. Real-space diffusion, trapping, and relaxation processes of photogenerated carriers in low-dimensional semiconductors are of interest from the viewpoint of fundamental physics as well as with regard to potential device applications. Moreover, coherent control of electronic excitation of the quantum-confined system is of great importance because of its possible application in quantum-information processing such as quantum computation, as described above.

In this chapter, current progress in the instrumentation and measurements of NSOM and its application to imaging spectroscopy of single-quantum constituents are described. The most critical element in NSOM is an aperture probe, which is a tapered and metal-coated optical fiber. We examine the design and fabrication of the probe with regard to aperture quality and the efficiency of light propagation. The recent dramatic improvements in spatial resolution and optical throughput are illustrated by single quantum-dot spectroscopy, which reveals the intrinsic nature of quantum-confined systems. Beyond such an application, real-space mapping of exciton wavefunctions confined in a quantum dot is also demonstrated.

2 General Description of NSOM

When a small object is illuminated, its fine structures with high spatial frequency generate a localized field that decays exponentially normal to the object. This evanescent field on the tiny substructure can be used as a local source of light illuminating and scanning a sample surface so close that the light interacts with the sample without diffracting. There are two methods by which a localized optical field suitable for NSOM can be generated. As illustrated in Fig. 1a, one method uses a small aperture at the apex of a tapered optical fiber coated with metal. Light sent down the fiber probe and through the aperture illuminates a small area on the sample surface. The fundamental spatial resolution of so-called aperture NSOM is determined by the diameter of the aperture, which ranges from 10 to 100 nm [10].

In the other method, called apertureless (or scattering) NSOM and illustrated in Fig. 1b, a strongly confined optical field is created by external illumination at the apex of a sharpened metal or dielectric tip [12]. Spatial resolution approaching the atomic scale is expected, and laboratory experiments have yielded resolutions ranging from 1 to 20 nm. A rather large (diffraction-limited) laser spot focused on a tip apex frequently causes an intense background that reduces the signal-to-noise ratio. This contrasts with what is done in aperture NSOM, where the aperture serves as a localized light source without any background. The general applicability of the apertureless method to a wide range of samples is currently being investigated.

The simplest setup for aperture NSOM, a configuration with local illumination and local collection of light through an aperture, is illustrated in Fig. 1c. The probe quality and the regulation system for tip–sample feedback are critical to NSOM performance, and most NSOMs use a method similar to that used in an atomic force microscope (AFM), called shear-force feedback, the regulation range of which is 0–20 nm. The light emitted by the aperture interacts with the sample locally. It can be absorbed, scattered, or phase-shifted, or it can excite fluorescence. Which of these occur(s) depends

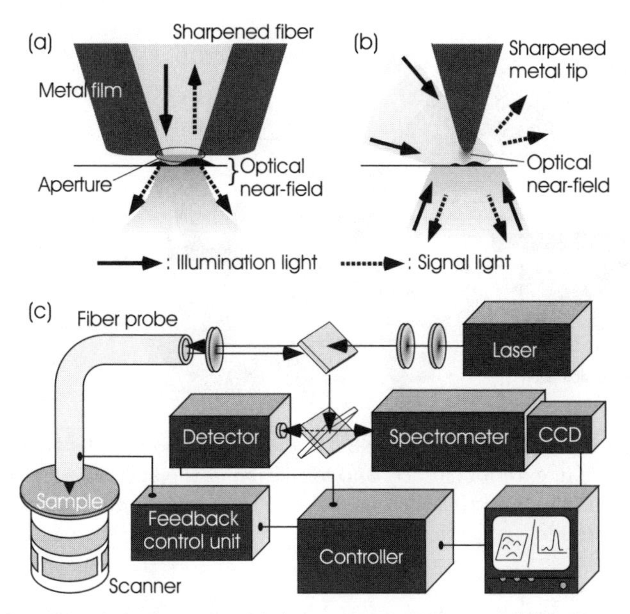

Fig. 1a-c. Schematic illustration of (**a**) aperture NSOM and (**b**) apertureless (scattering) NSOM. (**c**) Schematic of standard NSOM setup with a local illumination and local collection configuration

on the sample and produce(s) contrast in the optical images. In any case, light emerging from the interaction volume must be collected as efficiently as possible. When the sample is prepared on a transparent substrate, signal light is frequently collected with an objective lens arranged in a transmission configuration.

3 Design, Fabrication, and Evaluation of NSOM Aperture Probes

3.1 Basic Process of Aperture-Probe Fabrication

Great efforts have been devoted to the fabrication of the aperture probe, which is the heart of NSOM. Since the quality of the probe determines the spatial resolution and sensitivity of the measurements, tip fabrication remains of major interest in the development of NSOM. The fabrication of fiber-based optical probes can be divided into the three main steps illustrated in Fig. 2: (a) the creation of a tapered structure with a sharp apex, (b) the coating with a metal (Al, Au, Ag) to obtain an entirely opaque film on the probe, and (c) the formation of a small aperture at the apex.

There are two methods used to make tapered optical fibers with a sharp tip and reasonable cone angle. One is the heating-and-pulling method, where the fiber is locally heated using a CO_2 laser and is subsequently pulled

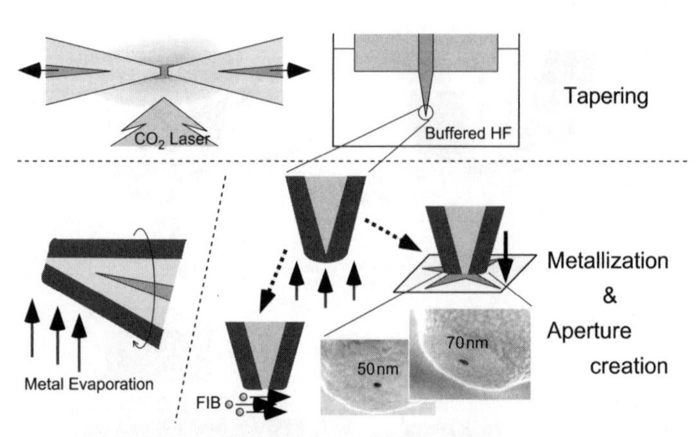

Fig. 2. Fabrication of an aperture NSOM probe: creation of the taper structure, metallization of an exterior surface, and formation of the aperture at the apex. The *inset* shows scanning electron micrographs of 50-nm and 70-nm apertures made by the impact method

apart [17]. The other method, based on chemical etching in a hydrofluoric acid (HF) solution, is a more easily reproducible production and can be used to make many probes at the same time [18,19]. To realize the ideal taper structures of the sort that will be discussed below, the chemical-etching method is advantageous because the taper angle can be adjusted by changing the composition of a buffered HF solution. Another important advantage of the chemical-etching method is the excellent stability of the polarization state of the probe. When the heating-and-pulling method is used, in contrast, temporal fluctuation of polarization occurs due to the relaxation of strain induced by the production process.

The technique most often used to form a small aperture is based on the geometrical shadowing method used in the evaporation of metal. Since the metal is evaporated at an angle from behind, the coating rate at the apex is much slower than that on the sides. The evaporated metal film generally has a grainy texture, resulting in an irregularly shaped aperture with asymmetric polarization behavior. The grains also increase the distance between the aperture and the sample, not only degrading resolution but also reducing the intensity of the local excitation. A method for making a high-definition aperture probe by milling and polishing using a focused ion beam has been developed [20], as has a simple method based on the mechanical impact of the tip on a suitable surface [21]. In both methods, the resulting probe has a flat end and a well-defined circular aperture. Furthermore, the impact method assures that the aperture plane is strictly parallel to the sample surface, which is important in minimizing the distance between the aperture and the sample surface.

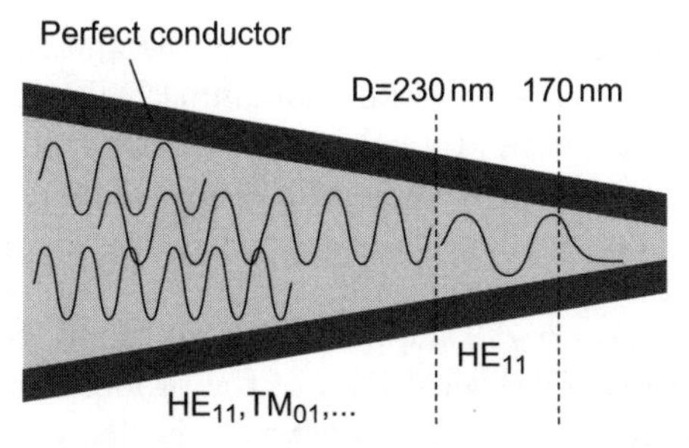

Fig. 3. Mode propagation in a tapered optical fiber probe coated with a perfect conductor at a wavelength of 633 nm

3.2 Tapered Structure and Optical Throughput

Improvement of the optical transmission efficiency (throughput) and collection efficiency of aperture probes is the most important issue to be addressed for the application of NSOM in the spectroscopic studies of nanostructures. In addition to contributing to the high sensitivity of NSOM measurements, optical probes with high throughput will open up attractive new areas of research, such as nonlinear processes and optical manufacturing on the nanometer scale. It is therefore of great importance to understand the limitations and possibilities of aperture probes with respect to light transmission.

The tapered region of the aperture probe operates as a metal-clad optical waveguide. The mode structure in a metallic waveguide is completely different from that in an unperturbed fiber and is characterized by the cutoff diameter and absorption coefficient of the cladding metal. Theoretical and systematic experimental studies have confirmed that the transmission efficiency of the propagating mode decreases in the region where the core diameter is smaller than half the wavelength of the light in the core (Fig. 3) [22]. The power that is actually delivered to the aperture depends on the distance between the aperture plane and the plane in which the probe diameter is equal to the cutoff diameter, a distance that is determined by the cone angle of the taper. We therefore proposed a double-tapered structure with a large cone angle [22]. This structure is easily realized using a multistep chemical-etching technique, as will be described below. With this technique, the transmission efficiency is much improved by two orders of magnitude as compared to the single-tapered probe with a small cone angle [21,22].

3.3 Simulation-Based Design of a Tapered Structure

Further optimization of the tapered structure is needed to achieve much higher probe efficiency. However, it is very time consuming to assess many structure parameters, such as the cone angle and taper length, by trial and error. Numerical analysis is a more reasonable way to attain an optimized structure efficiently and to understand the electromagnetic field distribution in a tapered waveguide including the vicinity of the aperture. Computational calculation by the finite-difference time-domain (FDTD) method is the most popular and promising method available for this purpose, because it can be easily applied to actual three-dimensional problems [23]. Although there have been many simulations focusing on the electric-field distribution in the vicinity of the aperture to examine the spatial resolution of NSOM, no calculations have been reported that deal with the light propagation in the tapered region in terms of the sensitivity of the probe. Here, using the three-dimensional FDTD method, we demonstrate the high collection efficiency of double-tapered probes including guiding optical fibers, as compared with single-tapered probes [24]. We also describe in detail the dependence of the collection efficiency on the cone angle and taper length.

Figure 4 illustrates the cross-sectional view of the FDTD geometry of the three-dimensional problem, which reproduces the experimental situation of single quantum-dot imaging. A fiber probe with a double- or single-tapered structure collects luminescence ($\lambda = 1\ \mu$m) from a quantum dot buried $3\lambda/40$ beneath the semiconductor (GaAs; $n = 3.5$) surface. We assume that the source of luminescence is a point-like dipole current linearly polarized along the x direction. The radiation caught by the aperture with a diameter of $\lambda/5$ propagates in the tapered region clad with a perfectly conducting metal and then is guided to the optical fiber waveguide. The refractive indices of the core and cladding of the fiber are 1.487 and 1.450, respectively. The intensity of the collected signal, I_{coll}, is evaluated by two-dimensionally integrating the electric field intensity in the core area of the optical fiber. The simulation box consists of a $120 \times 120 \times 360$ grid in the x, y, and z directions; the

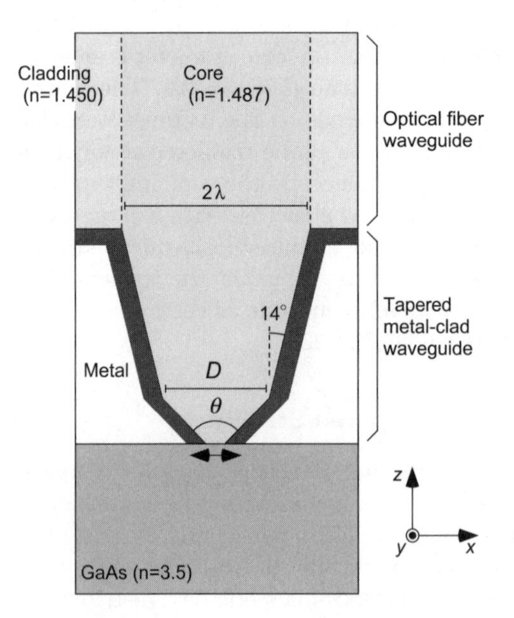

Fig. 4. Cross-sectional view of the FDTD geometry of the three-dimensional NSOM model

space increment is $\lambda/40$. We run the simulation employing Mur's boundary condition.

To demonstrate the performance of the double-tapered probe, we performed calculations for three types of probes, as shown in Fig. 5, where the spatial distribution of the electric-field intensity is shown on a logarithmic scale. In Fig. 5a and b, I_{coll} is compared for probes with $\theta = 28°$ and $\theta = 90°$. The I_{coll} ratio is estimated to be 1:32. Such a distinct improvement in I_{coll} can be attributed to the difference in the length of the cutoff region. By making the cone angle large and shortening the cutoff region, we can direct much of the radiation power towards the tapered region. Figure 5c shows the cal-

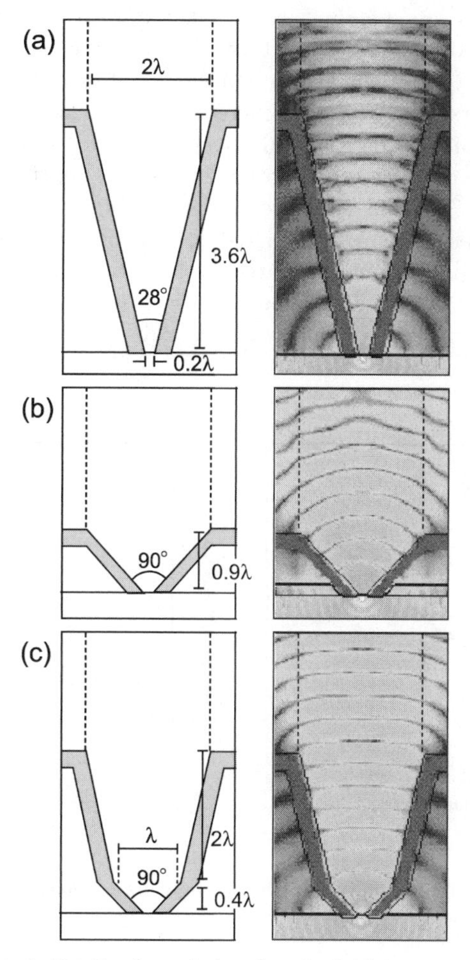

Fig. 5a-c. Calculated distribution of the electric-field intensity for three types of probes. (**a**) Single-tapered probe with a cone angle $\theta = 28°$, (**b**) single-tapered probe with $\theta = 90°$, and (**c**) double-tapered probe with $\theta = 90°$ and neck diameter $D = \lambda$

culation result in the case of a double-tapered probe whose cone angle is the same as in Fig. 5b. The neck diameter D is assumed to be λ, which is three times the cutoff diameter of the cylindrical waveguide clad with a perfectly conducting metal. I_{coll} of Fig. 5c is found to be three times greater than I_{coll} in Fig. 5b. The radiation pattern in Fig. 5c clearly illustrates that the second tapered region modifies the wavefront of the propagating light to match the guiding mode of the optical fiber, while the spherical-wave-like propagation in Fig. 5b cannot be coupled to the guiding mode so efficiently. To summarize, the collection efficiency of the double-tapered probe in Fig. 5c is greater by two orders than that of the conventional single-tapered probe in Fig. 5a.

Although we have demonstrated the advantage of a double-tapered probe, its performance is also dependent on various structure parameters. In Fig. 6a and b, the values of I_{coll} as a function of cone angle θ and neck diameter D, respectively, are plotted. The enhancement of I_{coll} with the increase in θ is easy to understand. I_{coll} will increase monotonously as θ approaches 180°. In the case of a realistic metal aperture, however, a large θ will cause diminished spatial resolution due to the finite skin depth of the metal. The optimum value of θ should be chosen by balancing the collection efficiency with the spatial resolution. As depicted in Fig. 6b, the dependence of I_{coll} on D is found to be more complicated and seems to be less essential. One significant result is that a neck diameter D as small as $3d_{\mathrm{c}}/2$ is preferable compared with a diameter of $3d_{\mathrm{c}}$ to attain high efficiency in coupling to the guiding mode of the optical fiber.

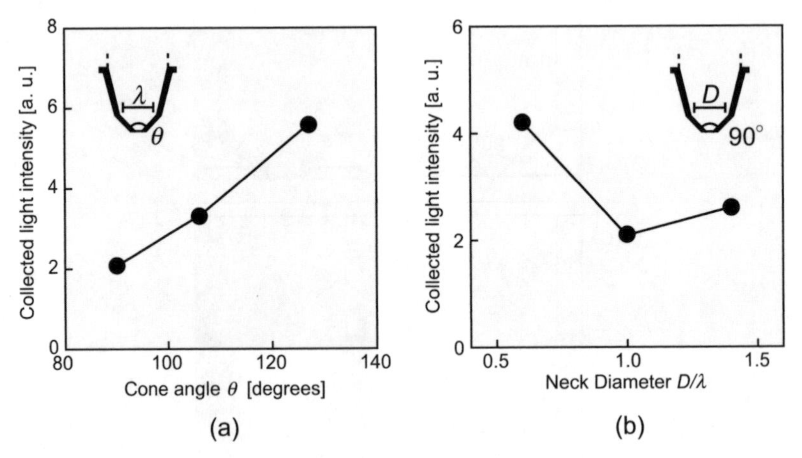

Fig. 6a,b. Plots of the intensity of collected light as a function of (**a**) cone angle θ and (**b**) neck diameter D

3.4 Fabrication of a Double-Tapered Aperture Probe

We used a chemical-etching process with buffered HF solution to fabricate the double-tapered probe. Since the details of probe fabrication with selective etching are described in [18], we describe the process only briefly here. The cone angle θ can be controlled by the buffering condition of the etching solution, which is adjusted by the volume ratio X of NH_4F maintaining the proportion of HF to H_2O at 1:1. Here, the composition of the solution is expressed as X:1:1. A two-step etching process is employed to make a double-tapered probe, as shown in Fig. 7a. In the first step, using a solution with a composition of $X = 1.8$, a short tip with a large cone angle of 150° is fabricated. Secondly, the guiding region is obtained with a solution of $X = 10$. As shown in the scanning electron micrograph of Fig. 7b, the resultant cone angle is approximately 90°. The neck diameter D can be controlled by the etching time in the second step.

The next step for metal coating and aperture formation is summarized in Fig. 8a. The entire exterior surface of the etched probe was metal coated with an Au film 200 nm in thickness using a sputtering coating method. A small aperture was created by pounding the metal-coated probe on a sapphire substrate or on the sample itself and squeezing the Au off to the side. Figure 8b shows a scanning electron micrograph of the 70-nm aperture, taken after conducting photoluminescence (PL) imaging several times. Not only was a smooth and flat end-face obtained in this fabrication process, but also

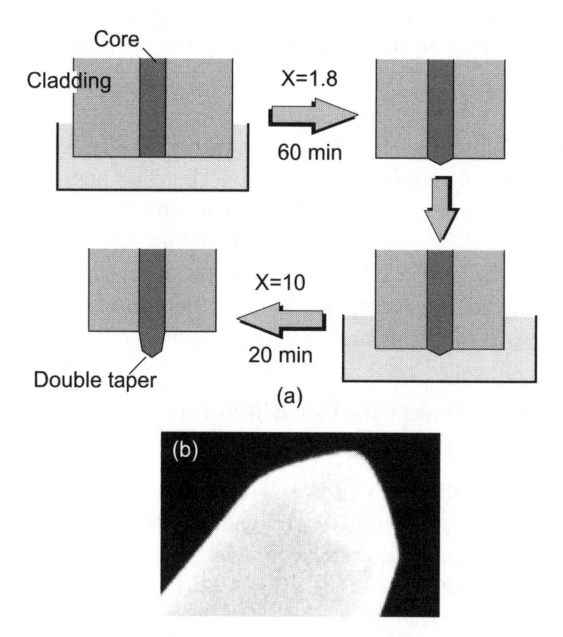

Fig. 7. (**a**) Two-step chemical etching for the fabrication of a double-tapered fiber probe. (**b**) Scanning electron micrograph of a double-tapered probe

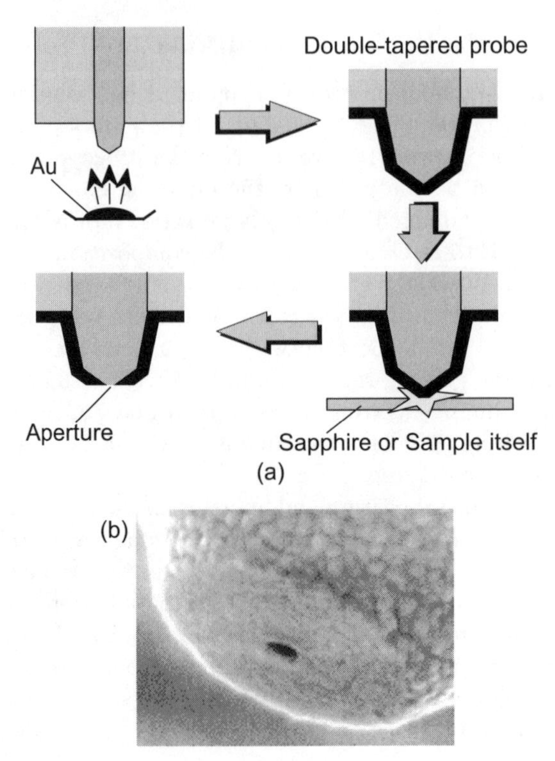

Fig. 8. (**a**) Process for coating the tapered probe with metal and formation of an aperture. (**b**) Scanning electron micrograph of the aperture with a diameter of 70 nm

a round and well-defined aperture. As mentioned above, the aperture plane is parallel to the sample surface to minimize the distance between the aperture and the sample surface. The size of the aperture can be selected by carefully monitoring the intensity of light transmitting from the apex, since the throughput of the probe is strictly dependent on the aperture diameter. Such controllability in aperture formation is demonstrated in the scanning electron micrographs of ultrasmall apertures in Fig. 9.

3.5 Evaluation of Transmission Efficiency and Collection Efficiency

To evaluate the transmission efficiency of aperture probes, researchers often measure the light power emitted by the aperture in the far-field region. However, in many experimental situations the instantaneous electric field is dominant in the interaction between light and matter. Thus, the far-field transmission coefficient (the ratio of the far-field light power emitted by the aperture to the power coupled into the fiber) does not reflect the field enhancement in the near-field region. However, in comparing probes with the

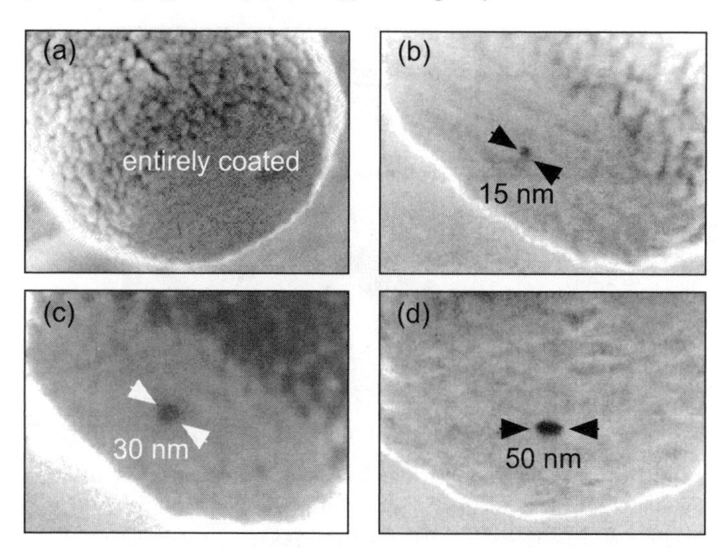

Fig. 9a-d. Scanning electron micrographs of (**a**) an entirely metal-coated probe and an apertured probe with aperture diameters of (**b**) 15 nm, (**c**) 30 nm, and (**d**) 50 nm

same aperture diameter in terms of the transmission efficiency in tapered regions, the far-field transmission coefficient can be valuable information.

In Fig. 10 the transmission coefficient is plotted against the aperture diameter for conventional single-tapered probes and for double-tapered probes with cone angles of 60° and 90°. As a light source for the measurement, we used a He-Ne laser ($\lambda = 633$ nm). It is clear that the transmission efficiency of the double-tapered structure with a large cone angle is improved by two orders of magnitude over that of the single-tapered probe with a small cone angle. In the case of a 100-nm aperture, the transmission coefficient is as large as 10^{-2}. The reduction of the throughput with a decrease of aperture diameter is not as severe as predicted by Bethe's theory, where the throughput must be proportional to d^{-6} (d: aperture diameter) [25]. Thus we obtain reasonable illumination intensity even in the case of a 10- to 30-nm aperture.

In addition to a high throughput for efficient illumination of light, superior collection of the locally emitted signal is critical for the observation of opaque materials such as semiconductors. We checked the collection efficiency of a 70-nm aperture probe quantitatively by collecting PL from an InGaAs QD buried 70 nm beneath the surface. We compared the PL intensity collected by the aperture probe with that collected simultaneously by an objective with a numerical aperture of 0.8 (Fig. 11 inset). In Fig. 11, the PL signal intensities are plotted as a function of distance between the aperture and the sample surface. Within 15 nm of the surface, the amount of light collected by the aperture probe increases rapidly and reaches a value as great as the amount collected by an objective with a numerical aperture of 0.8.

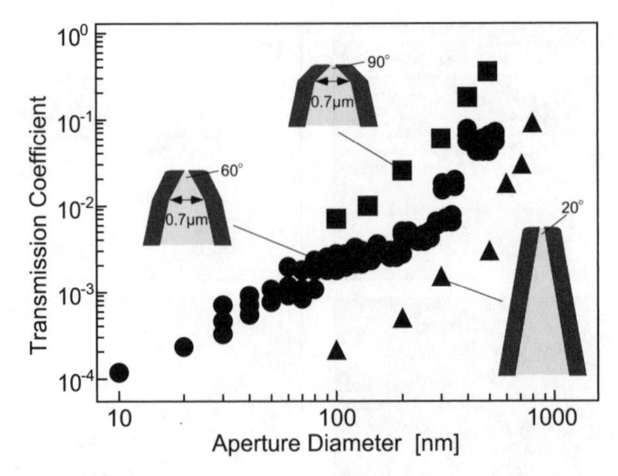

Fig. 10. Plots of the transmission coefficient of an aperture probe as a function of aperture diameter for single-tapered and double-tapered probes with cone angles of 60° and 90°. A He-Ne laser with a wavelength of 633 nm was used for this measurement

This suggests that fabrication of the flat-ended probe makes a considerable contribution to the efficient interaction of the aperture and evanescent field in the vicinity of the sample.

The overall throughput (or collection efficiency) of light is determined by various factors such as the wavelength of the propagating light and the corresponding dielectric constant of the cladding metal, as well as by the

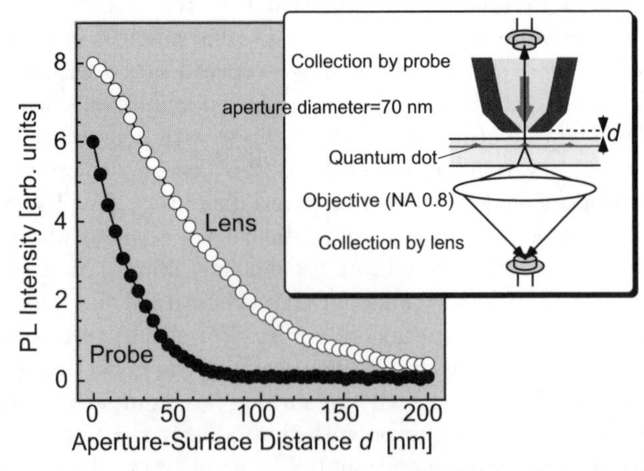

Fig. 11. Comparison of collection efficiency of an objective (*open circles*) and a probe with a 70-nm aperture (*closed circles*). Photoluminescence intensities collected from a single quantum dot are plotted as a function of distance between the aperture and the surface

structure of unperturbed fiber. The FDTD simulation of light propagation is a useful and efficient way to obtain the information needed to make an optimized structure.

3.6 Evaluation of Spatial Resolution with Single Quantum Dots

Fluorescence imaging of point-like emission sources is the most straightforward and reliable method for evaluating the spatial resolution of a probe. In particular, semiconductor quantum dots are most suitable in terms of their sizes, quantum efficiency, and optical stability. We evaluated the spatial resolution for various aperture diameters by PL imaging of single InAs self-assembled QDs [26]. The typical lateral size and height of the dots were 22 nm and 3 nm, respectively. The thickness of the GaAs cap layer, which determines the achievable spatial resolution, is of critical importance. While a thinner cap layer contributes to the higher resolution, it also causes the degradation of the optical quality of the QDs [4,27]. We used the QD sample with a cap layer of 20 nm thickness, taking into account the balance between the achievable spatial resolution and the optical quality.

By illuminating the sample with diode laser light ($\lambda = 685$ nm) through the aperture, we generated most of the carriers in a barrier layer surrounding the QDs. After diffusing in the barrier layer, the carriers were captured in the confined states of QDs. The PL signal from a single QD was collected by the same aperture. All of the measurements were performed at 9 K in a cryostat. The tapered structure, defined by a cone angle ($\theta = 90°$) and a neck diameter ($D = 1$ µm), was optimized from the viewpoint of both the optical throughput and the spatial resolution, based on the results of FDTD simulation.

Figure 12a–c shows the low-temperature PL images of InAs QDs obtained (center photon energy $E = 1.33$ eV, $\Delta E = 60$ meV) by the apertures with diameters of 135, 75, and 30 nm, respectively. The individual bright spots correspond to the PL signal from single QDs. The size of each spot becomes smaller with a decrease of the aperture diameter employed. To evaluate the spatial resolution in Fig. 12c, we plotted the cross-sectional profiles of PL signal intensity for the spots indicated by arrows in Fig. 12d. Full width at half-maximum (FWHM) of the PL signal profile was 37 ± 2 nm. Taking into account the size of QD (22 nm), we estimated the spatial resolution of this measurement to be about 30 nm, which corresponds to $\lambda/30$. In the illumination–collection mode operation of NSOM, the small aperture plays roles in both excitation and collection of the PL signal. In the case of carrier generation in the barrier layer, however, the excitation process does not contribute to such a high spatial resolution due to the carrier diffusion in several hundreds of nanometers. We conclude that high resolution could be achieved solely by the collection process of the PL signal.

We carried out the same measurements as described above for other probes and evaluated the PL spot size as a function of aperture diameter,

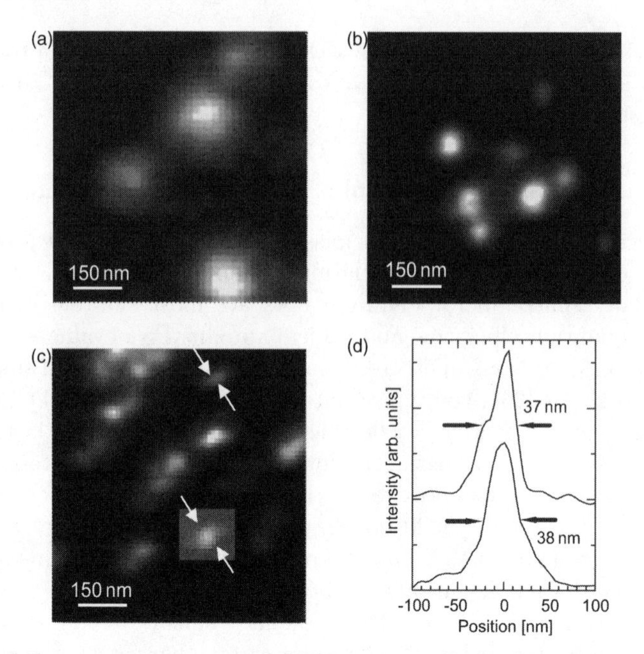

Fig. 12a-d. Low-emperature near-field PL images of InAs QDs, obtained by using the probes with aperture diameters of (**a**) 135, (**b**) 75, and (**c**) 30 nm, respectively. (**d**) Cross-sectional PL intensity profiles of the spots indicated by the *arrows*

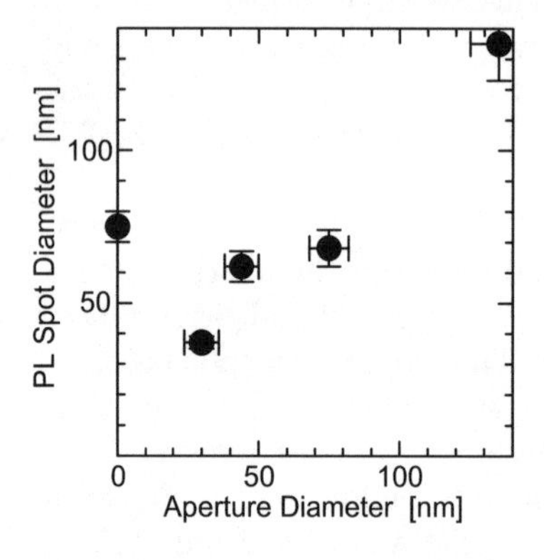

Fig. 13. PL spot diameter as a function of aperture size. The aperture size of 0 nm means that the probe does not have a physical aperture (entirely metal-coated probe)

as shown in Fig. 13. The aperture size of 0 nm means that there is no physical aperture at the apex, which was obtained by stopping the pounding procedure just before opening the aperture. In spite of the lack of physical aperture, we could detect a reasonable intensity of the PL signal and image with a resolution of 75 nm. Except for the 0-nm aperture, the spatial resolution is monotonously enhanced with a decrease of the aperture diameter. The spatial resolution is almost equal to the aperture diameter, whose behavior is also reproduced well by the FDTD simulation.

4 Super-Resolution in Single-Molecule Detection

NSOM fluorescence measurements are the simplest and most informative method because they provide high-contrast imaging of nanostructures, revealing the chemical composition and molecular structure as well as the defects and dopants in semiconductors. High-resolution optical imaging of biological samples with fluorescence labels is also a promising field of application. NSOM has the potential to create images of the distribution of such labels down to the level of single fluorophores. Furthermore, the presence of the probe (metal aperture) alters the fluorescence lifetime and emission pattern, making it possible to control radiation in the immediate vicinity of the probe (i.e., locally within the nanoenvironment).

Although NSOM offers attractive possibilities with regard to the detection of single molecules, the current spatial resolution of 100 nm is not sufficient for application to actual biological samples. Imaging with a very small aperture (< 30 nm) has often been impossible because the transmission efficiency decreases drastically with aperture diameter. Moreover, even a small grain on the aperture plane is problematic because the consequent increase of the aperture–sample distance reduces the spatial resolution. As mentioned in the previous section, however, the development of tailored probes with a high-quality aperture has solved this problem.

We have made fluorescence images of single dye molecules by using an aperture significantly smaller than that usually employed [28]. Cy5.5 dye molecules were dispersed on a quartz substrate and illuminated by He-Ne laser light through an aperture made of gold. The fluorescence from a single molecule was collected by the same aperture or by an objective in transmission configuration. Figure 14a shows a typical fluorescence image of individual molecules, and Fig. 14b shows magnified views of some of the bright spots in Fig. 14a. The diameter of each spot was estimated to be 30 nm, which corresponds to the aperture diameter produced by using the impact method. Step-like digital bleaching, which occurred at the moment indicated by the arrow, confirms that we actually observed single molecules. The highest resolution we achieved by using an aperture with a diameter of approximately 10 nm is shown in Fig. 15. The diameter of the fluorescence spot was as small as 10 nm.

The spatial resolution of 10 nm achieved in the experiment is inconsistent with conventional NSOM resolution; the optical spot generated at the aperture should be larger than the physical aperture diameter because of the finite penetration of light into the coated metal. When we use a gold-coated probe with a cone angle of 90°, the spatial resolution (spot size) should be limited to 30 nm at $\lambda = 633$ nm even in the case of a 10-nm aperture. To examine this discrepancy, we used an FDTD calculation to take into account a realistic situation, including the complex dielectric function of the gold film. Figure 16a shows the model for calculation. The radiation ($\lambda = 633$ nm) from a point dipole current, oscillating in the x direction, propagates in the ta-

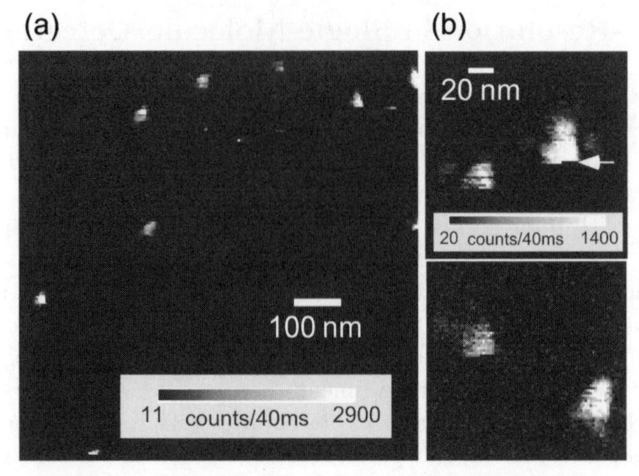

Fig. 14. (**a**) Typical near-field fluorescence image of single dye molecules. (**b**) Magnified view of one of the bright spots in part (**a**). Irreversible photobleaching occurred at the moment indicated by the *arrow*

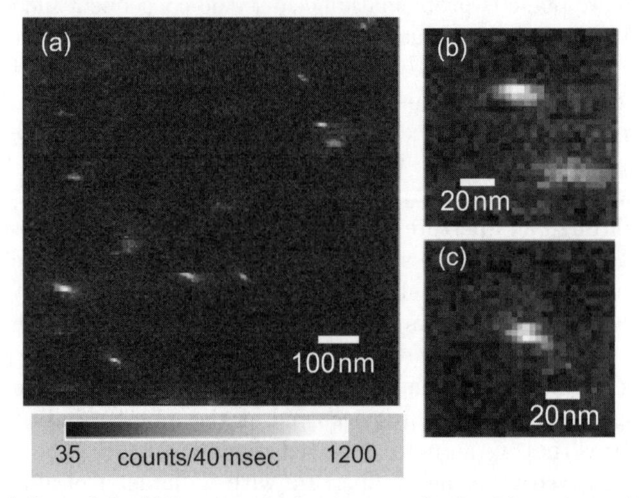

Fig. 15. (**a**) One of the highest-resolution images obtained by using a 10-nm aperture. (**b**) and (**c**) Magnified view of two of the bright spots in part (**a**)

pered region, and then a small light spot is created by the aperture. The cone angle is 90°, and the aperture diameter is 10 nm. The simulation box is 500nm $\times$ 500nm $\times$ 425nm in the x, y, and z directions, respectively, and the space increment is 2.5 nm. Figure 16b shows the cross-sectional profile of the electric-field intensity along the x direction 5 nm below the aperture. The twin-peak structure comes from the enhancement of the electric field at the edge of the aperture. The spot size, which is defined as a full width at half maximum of the profile, is estimated to be 16 nm. In Fig. 17, calculated

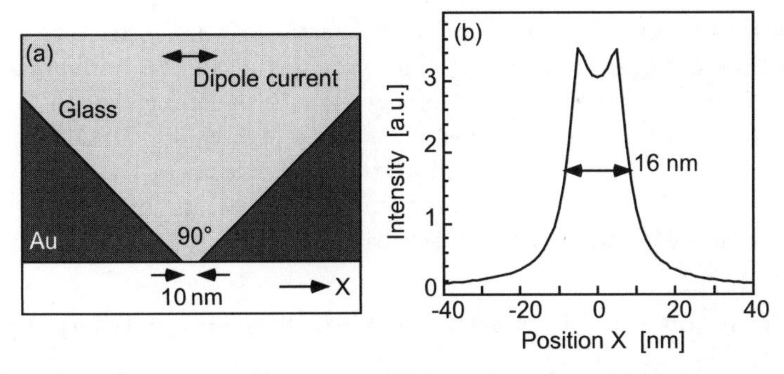

Fig. 16. (**a**) Simulation model for a gold-coated probe with a 10-nm aperture. (**b**) Cross-sectional profile of the electric-field intensity along the x direction 5 nm below the aperture

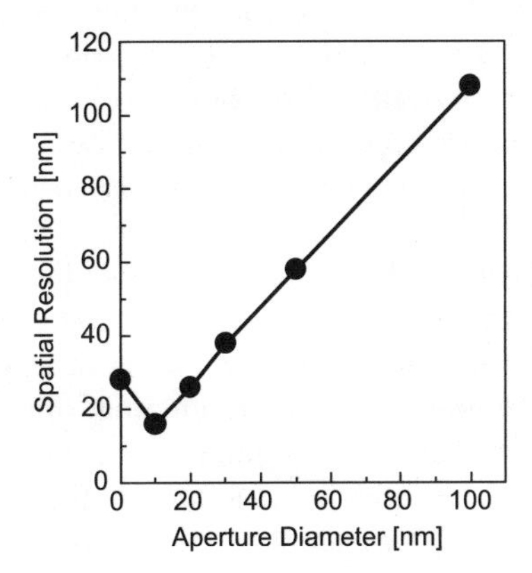

Fig. 17. Plot of calculated spot sizes (spatial resolution) as a function of the aperture diameter

spot sizes are plotted as a function of the aperture diameter. While we could expect to achieve 10-nm resolution from this simulation, a contribution of another mechanism, such as energy transfer from an excited molecule to the metal aperture, is also possible [29]. When the molecule is beneath the metal coating, the energy transfer shortens the fluorescence decay time and thus reduces emission intensity dramatically. On the other hand, no energy is transferred when the molecule is below the aperture hollow. Because the interaction distance in the process of energy transfer is only a few nanometers, we expect to reach a spatial resolution of less than 10 nm.

5 Single Quantum-Dot Spectroscopy

With the recent progress in the nanostructuring of semiconductor materials and in the applications of these nanostructured materials in optoelectronics, near-field optical microscopy and spectroscopy have become important tools for determining the local optical properties of these structures. In single quantum-dot spectroscopy, NSOM provides access to individual QDs, an ensemble of which exhibits inhomogeneous broadening due to the wide intraensemble distribution of sizes and strains [4,5,30–35]. NSOM can thus elucidate the nature of QDs, including the narrow optical transition arising from the atom-like discrete density of states, and the line broadening due to the interaction with phonons and/or carriers.

5.1 Homogeneous Linewidth and Carrier–Phonon Interaction

The electronic and optical quality of semiconductor QDs has been improved remarkably by using the Stranski–Krastanow growth mode. Owing to the atom-like density of states caused by the three-dimensional nanoscale confinement, the improvement of laser device performance such as ultralow threshold current and ultrahigh gain characteristics is theoretically predicted. Not only in terms of the device applications but also because of its fundamental physics, the QD is an interesting system due to its novel electronic and optical properties. The linewidth of PL or the absorption spectrum is one of the most fundamental and important parameters of these properties. There are intrinsic and extrinsic mechanisms responsible for the spectral broadening in semiconductor QDs. The former is homogeneous linewidth broadening, which is determined by the dephasing processes in electronic systems. The electron–phonon interactions, lifetime broadening, and carrier–carrier interactions are considered to cause the homogeneous broadening. The latter is inhomogeneous linewidth broadening due to the fluctuation in size, shape, and strain distribution of QD ensembles. Such inhomogeneity leads to variation in the optical transition energy from dot to dot and prevents us from understanding its intrinsic properties.

From the viewpoint of device applications, the homogeneous linewidth at room temperature is of great interest. The several mechanisms for linewidth broadening at high temperature such as carrier–carrier interaction [36] and carrier–phonon interaction [37] have been investigated by theoretical consideration. Compared with cryogenic measurement, however, single-QD PL spectroscopy at room temperature is technically more difficult, since the PL intensity decreases by a few orders due to the escape of carriers to the nonradiative recombination paths and/or thermal excitation to the barrier layer. For this reason, researchers have not yet reported the detailed experimental results of homogeneous linewidth broadening or the comparison between the experimental and the theoretical calculation results at high temperature.

In our single-QD PL spectroscopy study performed over a wide temperature range from 10 to 300 K, we observed that the spectral homogeneous linewidth increases with temperature, and it finally reaches a level greater than 10 meV at 300 K. We also found a correlation between the homogeneous linewidth and the interlevel spacing energy. These measured properties can be explained by theoretical calculations considering the electron–longitudinal optical (LO) phonon and electron–longitudinal acoustic (LA) phonon interactions [38].

The samples investigated in this study were $In_{0.5}Ga_{0.5}As$ self-assembled QDs grown on GaAs (100) substrate fabricated by gas source molecular beam epitaxy [39]. The substrate was not rotated during growth of the QDs, to create a gradient in the QD density. There were 70-nm thick cap layers covering the QDs. In order to achieve isolation of a single QD, we measured the region over the wafer where the density of the QDs was from 10^8 to 10^9 cm^{-2}. The typical lateral size and height of the dots were 32 ± 6 and 8 ± 2 nm, respectively, as measured by means of an atomic force microscope.

The QD sample on the scanning stage was illuminated with a laser diode ($\lambda = 687$ nm) through an aperture of the fiber probe under shear-force feedback control. The PL measurements were performed in the illumination and collection hybrid mode to prevent a decrease in spatial resolution due to carrier diffusion. The collected PL signal was sent into a monochromator equipped with a cooled charge coupled device (CCD). The spectral resolution of this measurement was about 1.5 meV. The NSOM head was mounted in a temperature-controlled cryostat.

Figure 18a shows the ground-state emission of single QDs under weak excitation conditions at various temperatures. The excitation power density was kept at a sufficiently low level, where the average number of photoexcited electron–hole pairs (excitons) is less than one in the dot. Under these excitation conditions, the additional PL linewidth broadening due to carrier–carrier interaction does not occur [40,41]. As shown in Fig 18a, the PL spectrum is sharp at 8 K and the FWHM of the spectrum is 1.5 meV, which is restricted by the spectral resolution in our instrumental response. The real PL linewidth is thought to be much narrower than 1.5 meV, as reported by other groups. The PL spectrum slightly broadens at 96 K, and the line shape can be reproduced by the Lorentzian function indicated by the fitting line. The FWHM of the PL spectrum, which corresponds to the homogeneous linewidth, is about 4 meV. With a further increase in temperature up to 250 K, the PL linewidth reaches approximately 10 meV. To demonstrate the linewidth variation from dot to dot, we measured the PL spectra of three different QDs at 250 K, as shown in Fig. 18b. The FWHM varies from 8.5 to 11 meV, and this type of variation of the spectral linewidths has also been reported [42].

Figure 19 shows the temperature dependence of the PL linewidths of single QDs from 8 to 300 K. We measured the PL spectra of several different QDs at various temperatures and plotted the experimental PL linewidths as open circles. Below 60 K, the PL linewidths are less than 1.5 meV. With an increase in temperature, the PL linewidth gradually broadens and finally becomes larger than 10 meV at 300 K. We compared this experimental finding with the results obtained by theoretical calculation. Our theoretical approach is fundamentally based upon the Kadanoff–Baym–Keldysh nonequilibrium Green's function technique [43]. The carriers interacting with phonons have

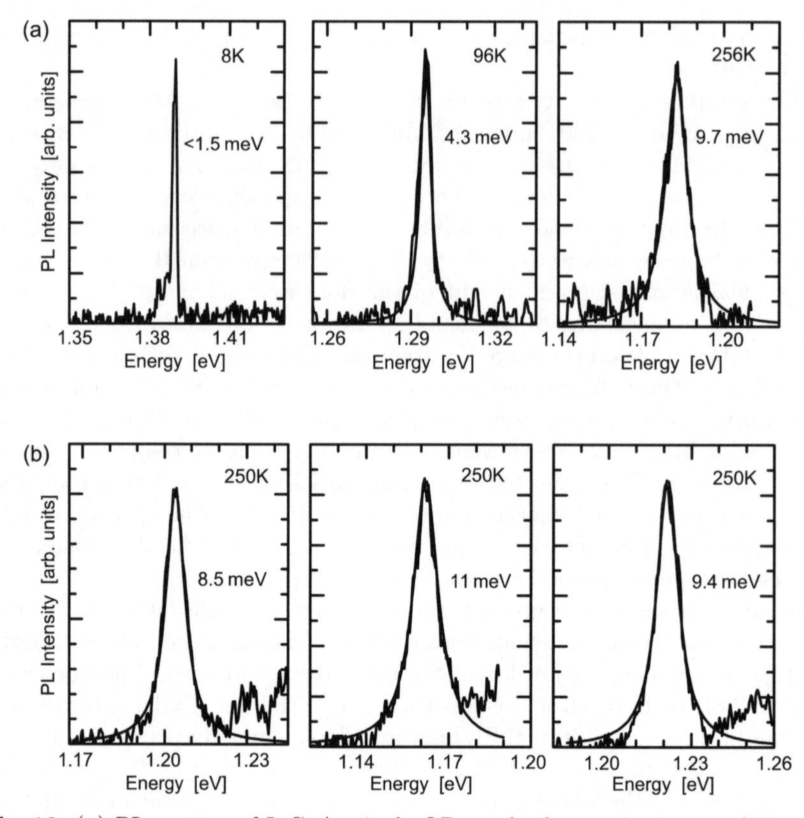

Fig. 18. (**a**) PL spectra of InGaAs single QDs under low excitation conditions at various temperatures. *Smooth lines* correspond to the Lorentzian function. (**b**) PL spectra of three different QDs at 250 K

a finite lifetime and violate a unique relation between the carrier energy and its wavevector. The spectral function is used to describe the relationship between them. To simplify the discussion, we consider only the lowest energy levels of the confined electron and the hole inside the quantum box. The LO phonon and LA phonon are assumed to be confined in the QD as well as the electron and the hole. We have plotted the calculation results in Fig. 21 as the solid squares and solid triangles, which correspond to the results considering both electron–LO and electron–LA phonon interactions and only the electron–LO phonon interaction, respectively. As a matter of course, the increase in the theoretical curves with temperature arises from the temperature dependence of the phonon numbers. In view of the fairly good agreement between the experimental findings and the results obtained by theoretical calculation, we believe that the homogeneous linewidths measured in this study are governed by the interactions between carriers and phonons. In addition,

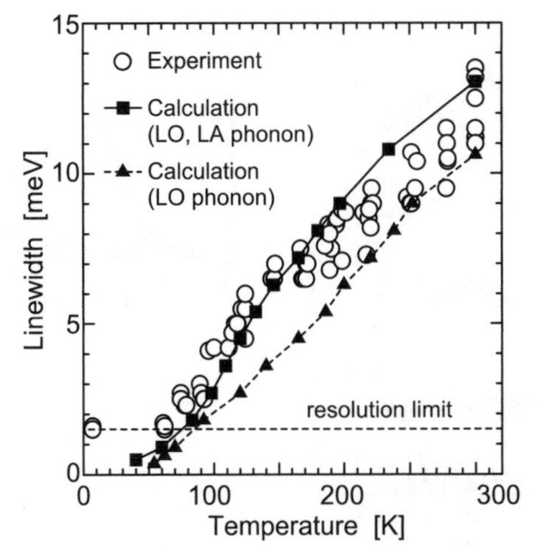

Fig. 19. Experimental and theoretical results for homogeneous linewidths as a function of temperature. *Solid squares* and *solid triangles* correspond to the calculated results considering both electron–LO phonon and electron–LA phonon interactions and only the electron–LO phonon interaction, respectively. The *dotted line* represents the spectral resolution in our measurement

we found that the interaction with LO phonons dominated the homogeneous linewidth in the high-temperature region.

Next, to understand why the homogeneous linewidths vary from dot to dot, we performed careful measurement of the homogeneous linewidths of many QDs at room temperature. To address the individual QD, we introduce the interlevel spacing energy (ΔE), which reflects the size of the QD. Figure 20a shows the excitation power dependence of the PL spectra of a single QD. The peak intensity of each spectrum is normalized by the excitation power. Under weak excitation conditions, a single peak is observed at 1.158 eV (E_1), which originates from the recombination of the ground state. When the excitation power is higher than 10 μW, another PL peak appears at the higher energy of 1.205 eV (E_2). From the threshold-like PL behavior as a function of excitation power, it is evident that this emission line of E_2 is associated with the first excited state. Figure 20b shows a plot of the PL spectrum at intermediate excitation power. Here, we define the interlevel spacing energy (ΔE) as the energy difference between the ground state and the first excited state ($= E_2 - E_1$).

Figure 21 shows the homogeneous linewidths as a function of ΔE, where the experimental data are plotted with open circles. The linewidths tend to increase with ΔE. Here, the absolute values and large variation (25–45 meV) of ΔE are considered to be reasonable, as determined by comparing these findings with the results obtained by theoretical calculation taking into account the averaged QD size and its distribution, respectively. Based upon the

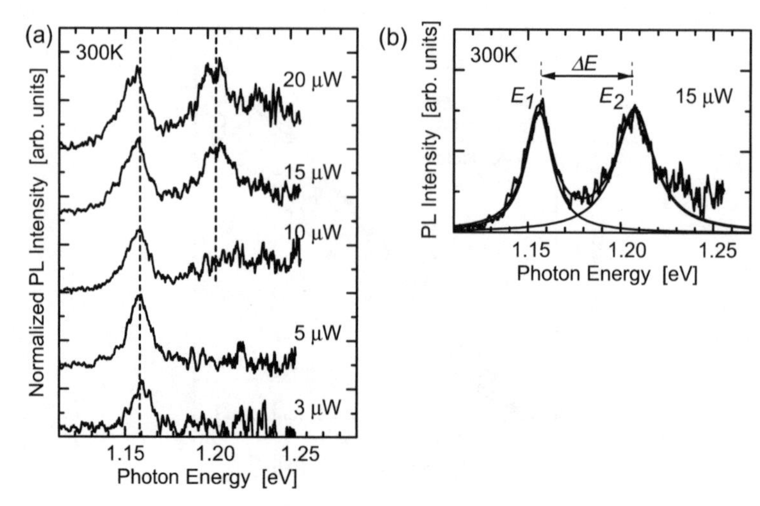

Fig. 20. (**a**) PL spectra of a single QD at various excitation powers ranging between 3 and 20 μW at 300 K. The *vertical axis* is normalized by the excitation intensity. (**b**) PL spectra of a single QD under intermediate excitation conditions, where several electron–hole pairs are excited in a QD. Interlevel spacing energy (ΔE) is defined as the energy difference between the ground-state emission energy (E_1) and the first-excited-state energy (E_2)

Green's function theory mentioned above, the homogeneous linewidths are computed for QDs of different size. The closed squares in Fig. 21 denote the calculated results considering the interactions with LO and LA phonons. The

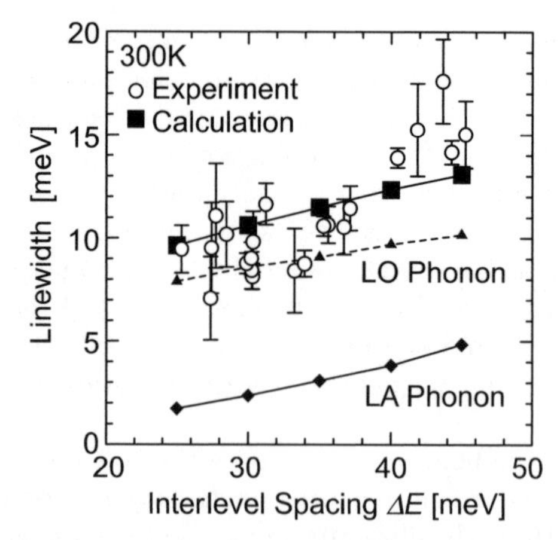

Fig. 21. Experimental (*open circles*) and theoretical (*closed squares*) results for homogeneous linewidth as a function of interlevel spacing energy (ΔE) at 300 K. Contributions from electron–LO phonon and electron–LA phonon interactions are also plotted separately

theoretical calculation results quantitatively reproduce the experimental results, indicating that the magnitude of electron–phonon interaction depends on the ΔE (size of QDs). That is, in the case of a smaller quantum dot, the overlap integral of the carrier and phonon eigenfunctions becomes larger, which leads to a more strongly coupled electron–phonon system. In Fig. 21, the contributions from LO– and LA–phonon interactions are also plotted separately. We can see that the ΔE dependence of the LA–phonon contribution is slightly larger than that of the LO–phonon contribution, which is due to the fact that the deformation potential includes the spatial derivative of the lattice displacement.

As for GaAs/$Al_{0.3}Ga_{0.7}As$ quantum-well structures, Gammon et al. [44] have reported the well-width dependence of the homogeneous linewidth. According to their experimental and theoretical results, the linewidth has a weak dependence on the quantum-well width. On the contrary, our experimental and theoretical results demonstrate that the strength of the electron–phonon interaction largely varies depending on the spatial confinement in a given QD. Such a size-dependent electron–phonon interaction implies that the QD would have a novel character with a strongly coupled electron–phonon mode.

5.2 Homogeneous Linewidth and Carrier–Carrier Interaction

Phenomena occurring under highly excited conditions have been one of the most significant topics that have been frequently studied in bulk, quantum-well (QW), and quantum-wire (QWR) systems [45–48]. In semiconductors occupied by many carriers, the carrier–carrier interaction appears as a consequence of a change of self-energy. Changes of the real and the imaginary part of a self-energy induce the shrinkage of a fundamental bandgap (bandgap renormalization) and an increase of the damping rate, respectively [45]. The experimental results show that the bandgap renormalization is reduced in accordance with the reduction of a dimensionality from bulk to QWR [48].

As mentioned in the previous subsection, the most essential problem in preventing the realization of ideal QD devices is the several kinds of inhomogenity of QD such as size, shape, and strain. However, the homogeneity of QD size is drastically improved when growth conditions are optimized. In this case, a narrow inhomogeneous broadening of 20 meV in the optical spectrum has been realized [49], and the performance of the QD lasers has been much improved. If the inhomogeneous broadening can be sufficiently suppressed, the intrinsic homogeneous linewidth becomes one of the restrictive parameters with respect to attaining the maximum gain. In the operation of a QD laser at room temperature, many carriers are injected in and around the QDs and the carrier–carrier interaction affects the homogeneous linewidth. Several scattering mechanisms via carrier–carrier interaction have been discussed, and the dephasing rate (linewidth) has been estimated by several theoretical calculations [37]. So far, however, the detailed experimental results involving carrier–carrier interaction in QDs at room temperature have not been reported.

In this subsection, we described a single-QD PL spectroscopy in high-excitation conditions at room temperature. We concentrated on the characteristic excitation-power-dependent shift and broadening of ground-state emission. These behaviors have been discussed while taking into consideration the carrier–carrier interaction in QDs.

Figure 22a shows the excitation-power dependence of PL spectra in a single QD and wetting layer (WL) at 300 K. In the weak-excitation condition, a single PL line originating from the recombination of an electron and hole pair in the lowest energy level (ground state: E_1) is observed at 1.222 eV. With an increase of the excitation power density, additional lines associated with excited states (E_2, E_3) appear and grow at 1.252 and 1.280 eV. The shape of each spectrum is well reproduced by a sum of Lorentzian functions. The precise values of the transition energy and the linewidth are determined from the fitting results. The additional broadening and spectral redshift of the ground-state emission are observed at several kW/cm^2. Such behaviors have been commonly observed in the twenty QDs that we have investigated.

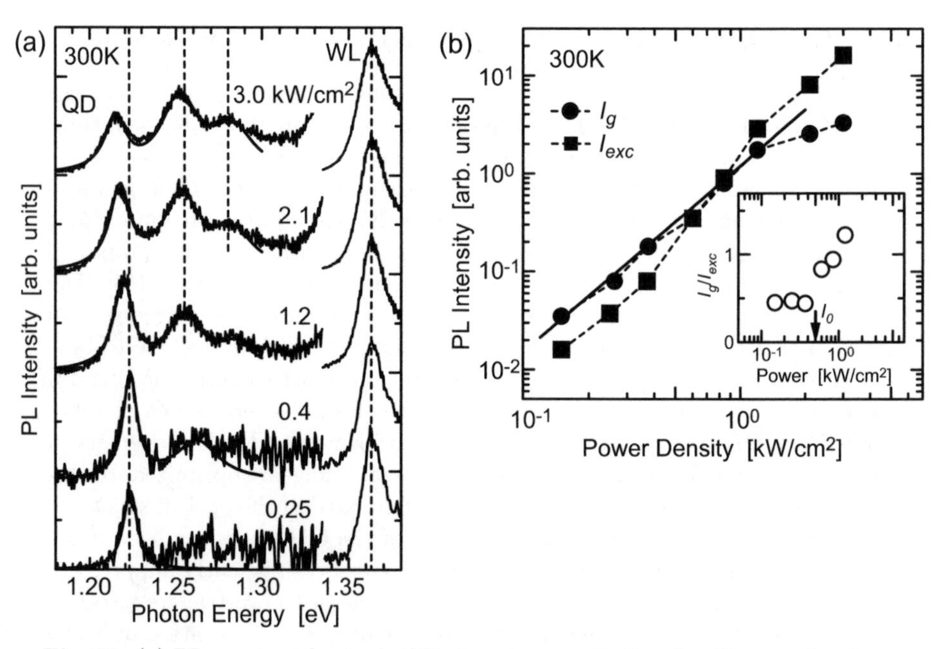

Fig. 22. (**a**) PL spectra of a single QD at various excitation densities ranging from 0.25 to 3.0 kW/cm^2. (**b**) Excitation power dependence of integrated PL intensity at ground-state emission (I_g: *closed circles*) and at the first excited state (I_{exc}: *closed squares*). *Inset* shows the PL intensity ratio (I_g/I_{exc}) as a function of excitation density. The excitation power density at which the PL intensity ratio exhibits a critical increase is defined as I_0

Figure 22b shows the integrated PL intensities of both the ground state and the first excited state as a function of excitation power densities. The PL intensity of the ground state (I_g: closed circles) grows linearly as a function of excitation power density and reaches saturation at approximately 1.5 kW/cm^2. The nonlinear power dependence of the PL intensity indicates that the nonradiative recombination process is dominant at room temperature [50]. The PL intensity of the first excited state (I_{exc}: closed squares) increases more rapidly than I_g above 0.5 kW/cm^2. The inset of Fig. 22b shows the excitation-power dependence of the ratio of the emission intensity from the first excited state to that from the ground state ($=I_g/I_{exc}$), which reflects the degree of carrier population in the first excited state. The PL intensity ratio is constant below 0.5 kW/cm^2. The finite population in the first excited state at the lowest excitation condition is well understood by the thermal carrier excitation at 300 K. The ratio exhibits critical increases above 0.5 kW/cm^2 because additional carriers occupy the first excited state due to the strong excitation. Hereafter, we define the power density where the PL intensity ratio critically changes as I_0.

To study the many-body effect in the QD, we have plotted the spectral energy shift of the ground state (E_1) and the first excited state emission (E_2) of the QD and WL as a function of excitation power in Fig. 23. Both E_1 and E_2 shift to the lower energy side with an increase in excitation density [51]. As the PL peak energy of the WL and GaAs barrier layer (not shown) does not change, the redshift of E_1 and E_2 cannot be explained by simple bandgap shrinkage of the barrier layer; rather, it is caused by the bandgap renormalization of the QD itself due to the Coulomb interaction between carriers. When the excited carrier density is 2×10^{17} cm^{-3}, the value of bandgap renormalization is evaluated to be 8 meV, which is in good agreement with the theoretical predictions [52,53].

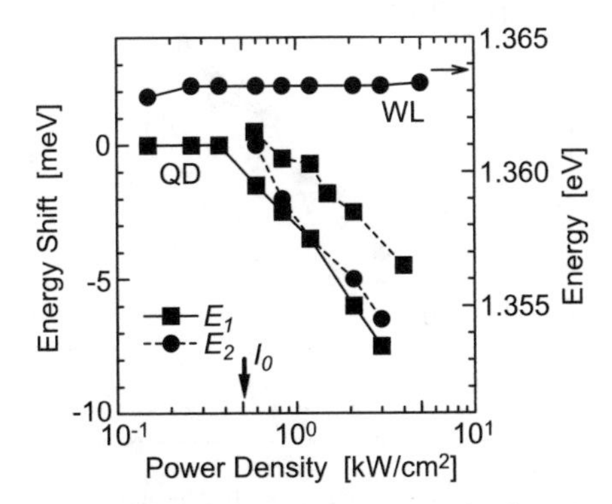

Fig. 23. Energy shifts of the ground-state emission (E_1) and the first excited state (E_2) of QD as a function of excitation power density. The PL peak energy of WL as a function of power density is also plotted

Next we examined additional broadening by evaluating the linewidth of the ground-state emission. Figure 24 shows a dependence of the emission linewidth (A: closed squares) on the excitation power density. Results of two different QDs (B: open circles, C: open squares) are also plotted. The linewidth remains constant below I_0. In this regime, the broadening is well explained by the dephasing process due to the interaction between electron (carrier) and phonon, which has already been discussed in detail [38]. At higher excitation, the linewidth increases with the excitation power. Through the investigation of many QDs we found that the values of I_0 were scattered from QD to QD, and we also found that additional broadening always starts to occur at I_0, that is, when the carriers are populated in the excited state of the QD. The strong dependence of this broadening on the excitation power implies that the interaction between carriers, such as the Auger scattering process, would lead to an increased dephasing rate. We should consider two types of Auger scattering process in the QD system grown in the S–K mode. One is the scattering process that takes place between confined carriers in the QD and free carriers in the two-dimensional (2D) WL. The other is the intra-QD process, in which carriers in the QD interact each other.

We will discuss which processes are dominant in terms of causing the excitation-power-dependent broadening in the QD system. First we consider the scattering process between confined carriers in a QD and free carriers in a 2D WL. A theoretical calculation predicts that the exchange process between a carrier in the QD and a free carrier in the WL causes a broadening of several meV at high carrier density and that the value of broadening is determined by the carrier density in the WL. However, this mechanism should be ruled out because the broadening behavior in the experiment differs from

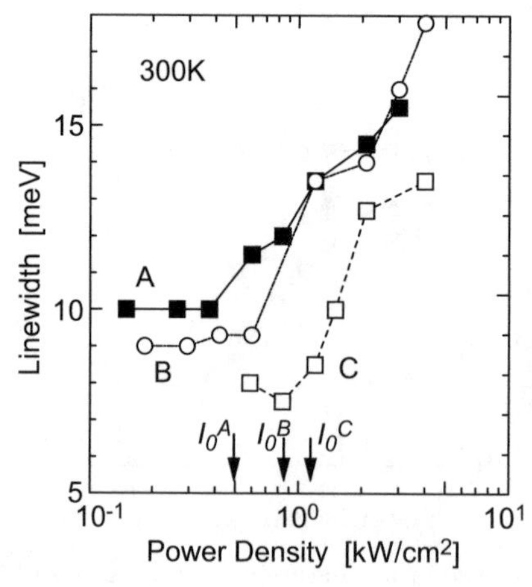

Fig. 24. Linewidth of the ground-state emission (E_1) as a function of excitation power density for three different QDs

QD to QD. (Through the 2D mapping of PL intensity, we confirmed that the carrier density in WL was uniform in the observation area.)

Next, we consider the intra-QD scattering process, where the broadening depends on the carrier population in the QD [54,55]. The correlation between the additional broadening and the carrier population in the excited states of the QD, as shown in Fig. 24, indicates that the intra-QD process is a dominant mechanism for the broadening. From the experimental results, we conclude that an Auger scattering process with the following scenario is most probable. First, there is a relaxation of the hole into the QD from the WL and the excitation of an electron into a higher level via Coulomb interaction. As a next step, the electron relaxes to the lowest level and the hole is emitted into the WL. Theoretically, the additional broadening of several meV is possible if the capture and emission processes are efficient. Further theoretical studies that take the realistic electronic structure into consideration will be needed for the quantitative explanation of the experimental results.

6 Real-Space Mapping of Exciton Wavefunction Confined in a QD

As described in the previous section, the near-field optical characterization of semiconductor quantum structures, with a resolution of approximately 0.2 μm, reveals sharp PL spectra from individual quantum constituents and provides a significant contribution to the investigation of their intrinsic electronic structure and dynamics. Beyond such single-constituent spectroscopy, the next challenge of near-field optical measurement is to illustrate directly the internal features of quantum-confined systems [56]. If the spatial resolution of NSOM reaches the length scale of the quantum structure, theoretical studies predict that the NSOM will allow mapping of the real-space distribution of eigenstates (wavefunctions) within quantum-confined systems [57]. Moreover, in combination with ultrafast pulsed excitation, the local access to single-QD systems enables control of individual eigenstates in a coherent manner, which provides a fundamental technique in quantum-information technology.

In this section, the near-field optical mapping of exciton and biexciton center-of-mass wavefunctions confined in 100-nm GaAs interface QDs is demonstrated. The QDs investigated in this study are much smaller than those investigated in a previous report [56]. The successful imaging of such a small quantum-electronic structure is attributed to improvement in the design and fabrication of the near-field fiber probe.

The sample investigated in this study was a 5-nm (18–19 ML) GaAs QW sandwiched between layers of AlAs and $Al_{0.3}Ga_{0.7}As$ grown by molecular-beam epitaxy. Growth interruptions for 2 min at both interfaces led to the formation of naturally occurring QDs, as shown in Fig. 25. The GaAs QW was capped with a 15-nm $Al_{0.3}Ga_{0.7}As$ layer and a 5-nm GaAs layer. As

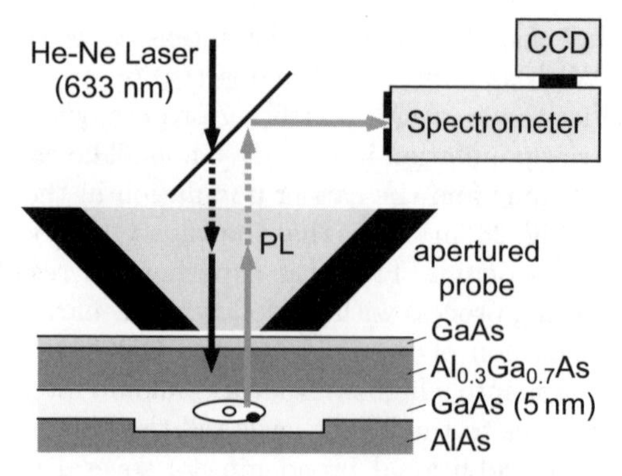

Fig. 25. Schematic of the sample structure and the configuration of the near-field optical measurement

we see in the above section, a spatial resolution as high as 30 nm can be achieved even when QDs are covered with 20-nm cap layers. As depicted in Fig. 25, the sample was nonresonantly excited with He-Ne laser light ($\lambda =$ 633 nm) through the aperture and the resultant PL signal was collected via the same aperture to prevent a reduction of the spatial resolution due to carrier diffusion. The near-field PL spectra were measured at every 10-nm step in a 210nm $\times$ 210nm area, and the near-field images were constructed from a series of PL spectra.

Figure 26a shows near-field PL spectra of a single QD at 9 K at excitation densities ranging from 0.17 to 3.8 μW. A single sharp emission line (denoted by X) at 1.6088 eV is observed below 0.4 μW. With an increase in excitation power density, another emission line appears at 1.6057 eV (denoted by XX). In order to clarify the origin of these emissions, we show the excitation power

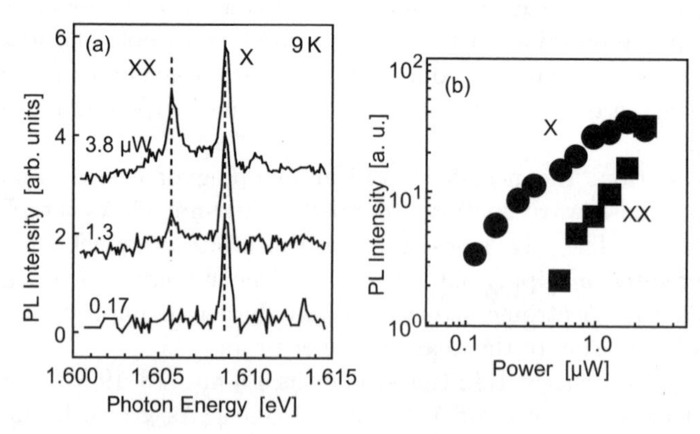

Fig. 26. (**a**) Near-field PL spectra of a single QD at 9 K under various excitation power densities. (**b**) Excitation power dependence of PL intensities of the X and the XX emission line

dependence of PL intensities in Fig. 26b. Based on the linear increase of the intensity of the X emission line and the saturation behavior, the X line is identified as an emission from a single-exciton state. The quadratic dependence of the XX emission with excitation power density indicates that XX is attributed to an emission from a biexciton state [47,58]. The identification of the XX line is also supported by the correlation energy of 3.1 meV (the energy difference between the XX line and the X line), which agrees well with the value reported previously [47,58].

High-resolution PL images in Fig. 27a and b were obtained by mapping the PL intensity with respect to the X and the XX lines under the excitation density of 1.3 W/cm^2. In this excitation regime, both the emissions from single-exciton (X) and biexciton (XX) states appear in a PL spectrum. The two images were obtained by selecting the detection peak in a series of PL spectra. The exciton PL image in Fig. 27a manifests an elongation along the [−110] crystal axis. The image sizes along the [110] and [−110] axes are 80 nm and 130 nm, respectively, which are larger than the PL collection spot diameter, that is, the spatial resolution of the NSOM. The size and the elongation along the [−110] axis are consistent with a previous observation made with a scanning tunneling microscope [5]. These findings lead to a conclusion that the local optical probing of the NSOM directly maps out the center-of-mass wavefunction of an exciton confined in a monolayer-high island elongated along the [110] crystal axis.

We also obtained the elongated biexciton PL image along the [−110] crystal axis and found a clear difference in the spatial distribution between the exciton and the biexciton emissions. This tendency was reproduced for other QDs in the same measurement. The different properties between exciton and biexciton wavefunctions may explain this. Since the relative-motion wavefunction of the biexciton state extends more widely than does that of the exciton state, the center-of-mass motion of the biexciton is more strongly confined in the monolayer-high island.

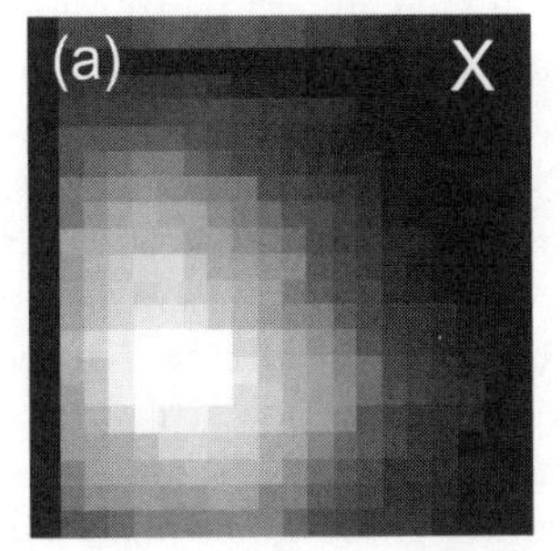

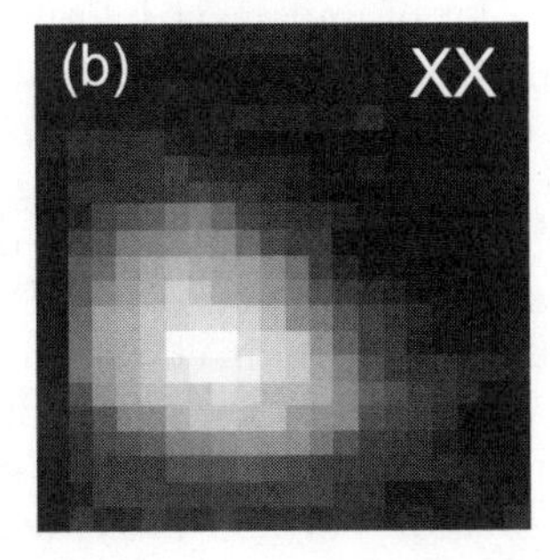

Fig. 27a,b. High-resolution PL mapping for the emissions from (**a**) the exciton state (X) and (**b**) the biexciton state (XX). The excitation power density and scanning area are 1.3 W/cm^2 and 210nm × 210nm, respectively

7 Carrier Localization in Cluster States in GaNAs

In contrast to the well-defined quantum-confined systems such as QDs grown in a self-assembled mode, the more common disordered systems with local potential fluctuations leave unanswered questions. For example, a large reduction of the fundamental bandgap in GaAs with small amounts of nitrogen is relevant to the clustering behavior of nitrogen atoms and resultant potential fluctuations. NSOM characterization with high spatial resolution can give us a lot of important information that is useful in our quest to fully understand such complicated systems, such as details about the localization and delocalization of carriers, which determine the optical properties in the vicinity of the bandgap.

The GaNAs and GaInNAs semiconductor alloy system has attracted a great deal of attention due to its interesting physical properties, such as its extremely large bandgap bowing [59], and its wide range of potential applications in solar cells and optoelectronic devices. In particular, for long-wavelength semiconductor laser application, high-temperature stability of the threshold current is realized in the GaInNAs/GaAs quantum well as compared to the conventional InGaAsP/InP quantum well due to strong electron confinement [60]. However, GaNAs and GaInNAs with a high nitrogen concentration of more than 1% have been successfully grown only under nonequilibrium conditions by molecular beam epitaxy [61,62] and metalorganic vapor phase epitaxy [63].

The incorporation of nitrogen generally induces degradation of optical properties [64–69]. To date, several groups of researchers have reported characteristic PL properties of GaNAs and GaInNAs, for example, the broad asymmetric PL spectra [64,67,69] and the anomalous temperature dependence of the PL peak energy [65,68]. With respect to this characteristic PL behavior, it has been pointed out in the realm of macroscopic optical characterization that carrier localization plays an important role. By contrast, microscopic optical characterization techniques with high spatial resolution can give us a lot of important information that allows us to understand the optical properties, especially the localization or delocalization of carriers.

In this section, we show the results of spatially resolved PL spectroscopy at both 7.5 and 300 K obtained using NSOM with a high spatial resolution of 150 nm. At low temperature, we observed sharp structured PL spectra and the spatial distribution of the PL intensity of the GaNAs alloy. This spatial inhomogeneity is direct evidence of carrier localization in the potential minimum case caused by the compositional fluctuation [70].

The samples investigated in this study were GaNAs epilayers grown on [001] GaAs substrates by low-pressure metalorganic vapor phase epitaxy. The details of the growth conditions are described in [71]. The GaNAs layer was sandwiched by a 100-nm thick GaAs buffer layer and a 50-nm thick GaAs cap layer. The GaNAs layer was 500 nm thick, and the N concentration was estimated to be 0.8% by high-resolution X-ray diffraction and secondary ion mass spectroscopy. After growth, thermal annealing was performed for 10 min in H_2 ambient at 700 K to improve the optical properties [71]. The macro-PL spectra were measured at both 10 and 300 K with excitation by means of a laser diode ($\lambda = 687$ nm). In the NSOM measurement, the sample was also illuminated by the laser diode through the aperture of the fiber probe, and the PL signal from the GaNAs epilayer was collected by the same aperture.

Figure 28a shows the excitation-power dependence of the macro-PL spectrum from a GaNAs epilayer at 10 K. Broad and asymmetric PL spectra with a low-energy tail as shown in Fig. 28a were observed under low-excitation conditions. With an increase in excitation power density, the PL intensity of the low-energy tail gradually reaches saturation, and the peak energy of the PL spectra (solid square) is shifted to the higher-energy side. In the PL spectrum under the highest-excitation conditions, we can see the appearance of a shoulder (open circles) corresponding to the transition energy of the band edge (1.343 eV) determined by the photoreflectance spectrum [72]. Figure 28b shows the excitation power dependence of the macro-PL spectra at room temperature. In contrast to what occurs at low temperature, the peak energy indicated by the open circles and the shape of the PL spectra do not change with an increase in excitation power. Here, it should be noted that we observed no broad emission due to recombination of the carriers at a deep trap in GaAs, and the quality of this sample was high. From the PL

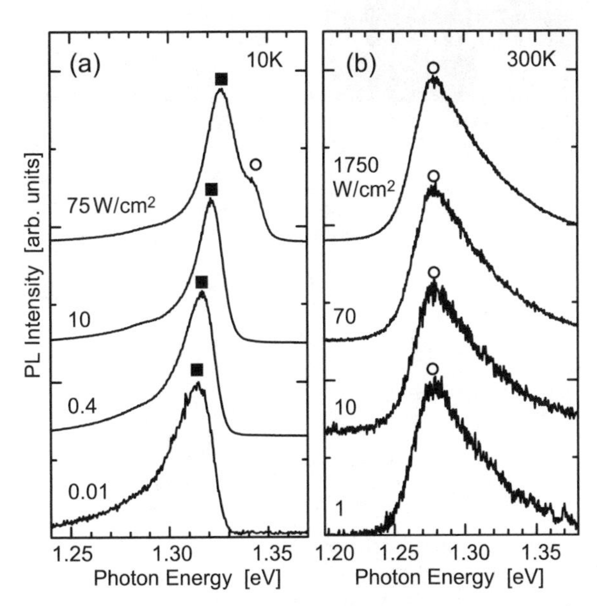

Fig. 28a,b. Power dependence of the macrophotoluminescence (PL) spectra of a GaNAs epilayer at (**a**) 10 and (**b**) 300 K. The *vertical* axis of the PL spectrum was normalized by the maximum value of each peak

behavior as a function of excitation power dependence, it is evident that the carrier-recombination mechanism drastically changes from 10 to 300 K.

In order to investigate the spatial distribution of the optical properties, we performed spatially resolved PL spectroscopy with NSOM at room temperature. Figure 29a shows a typical topographic image of the sample surface. The flatness of the sample surface is within a few nanometers, which does not affect the artificial effect for near-field optical images. Figure 29b shows the monochromatic PL image obtained at a detection energy of 1.268 eV with a spatial resolution of 150 nm at 300 K. Weak contrast on a large scale of a few micrometers is observed. Figure 29c shows the near-field and macro-PL spectra at 300 K. The near-field PL spectrum is almost the same as the macro-PL spectrum, indicating that the carriers do not localize spatially. The PL spectrum with a higher-energy tail is well reproduced by the fitted curves (open circles), assuming that the carriers are found in a bulk system with Boltzmann distribution. Therefore, we can conclude that the dominant emission

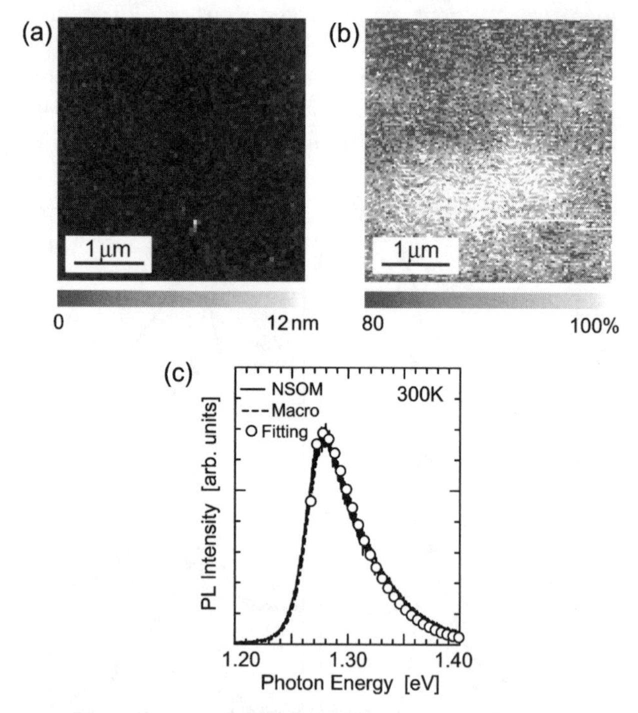

Fig. 29. (**a**) Topographic image of the sample surface. The scanning area is 4.0 μm × 4.0 μm. (**b**) Monochromatic PL images obtained at a detection energy of 1.268 eV with a spectral window of 6 meV at 300 K. (**c**) PL spectra obtained with the macro (*dotted line*) and near-field (*solid line*) configuration. The power density of the near-field PL measurement is 200 W/cm^2. The *open circles* represent the *fitted curves* assuming the free-carrier recombination at the band edge

of the GaNAs epilayer at room temperature comes from the recombination of the delocalized carriers at the band edge.

To examine the difference in the mechanism of recombination of carriers between room temperature and cryogenic temperature, we conducted a near-field PL measurement at 7.5 K. Figure 30a and b shows the low-temperature monochromatic PL images obtained at a detection energy of 1.316 and 1.328 eV, respectively. We observed high-contrast images with structures on a scale of within 1 µm, as shown in both Figs. 30a and 30b. By comparing Figs. 30a and 30b, we can see that the contrast pattern of the monochromatic image changes in accordance with the detected photon energy. These high-contrast and energy-dependent images suggest the existence of energy fluctuations on a scale of within 1 µm. Figure 30c shows the integrated PL intensity image obtained at a photon energy of 1.325 eV with a spectral window of 30 meV to detect the whole PL signal at 7.5 K. Spatial inhomogeneity of the integrated PL intensity with a scale of several hundred nanometers was observed, indicating that the density of the defect and the nonradiative recombination center are distributed spatially [73].

Figure 31a shows the near-field PL spectrum and its power dependence as determined by fixing the fiber probe at a certain position. Under the lowest excitation conditions (the lowest spectrum in Fig. 31a), several sharp peaks at around 1.32 eV, not seen in the macro-PL spectrum (dotted line), are clearly observed. With an increase in excitation density, the sharp PL peaks gradually show saturation and the center of gravity in the PL spectrum shifts to the higher-energy side. Finally, we observed structureless PL spectra similar to the macro-PL spectra. Judging from the sharp spectral features and the rapid saturation behavior, it seems that the sharp PL peaks observed under low-excitation conditions come from the recombination of localized carriers (excitons).

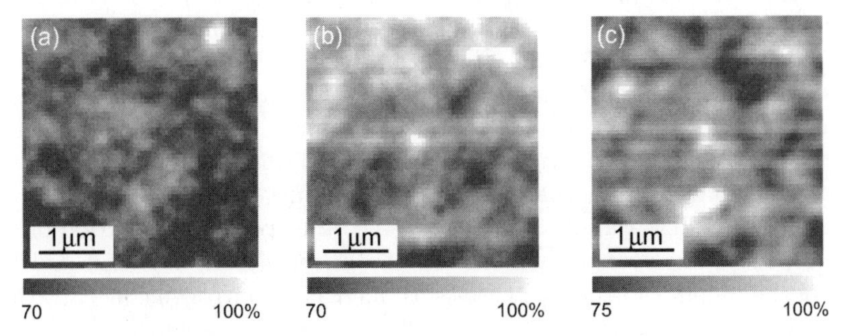

Fig. 30. (**a**), (**b**) Monochromatic PL images obtained at a photon energy of 1.316 and 1.328 eV, respectively, with a spectral resolution of 4 meV, measured at 7.5 K. (**c**) Integrated PL intensity image obtained at a photon energy of 1.325 eV with a spectral window of 30 meV. The scanning area and excitation power density are 4.0 µm×4.0 µm and about 5 W/cm^2, respectively

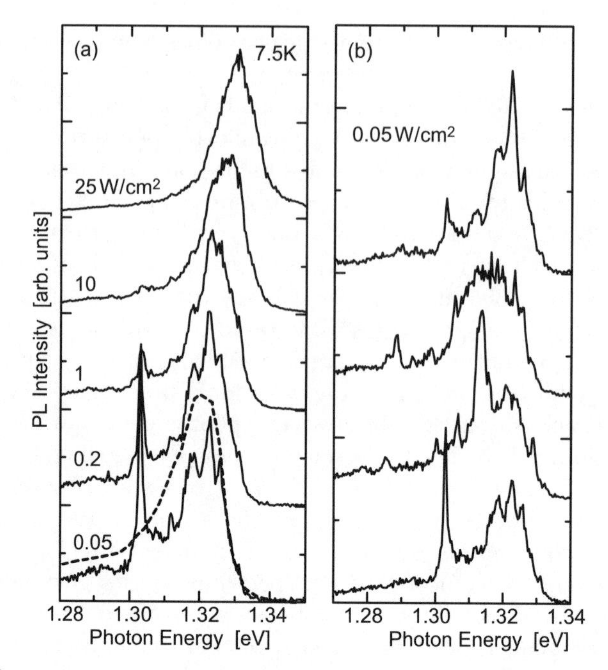

Fig. 31. (**a**) Near-field PL spectra at various excitation power densities at 7.5 K. The *dotted line* represents the macro-PL spectrum. (**b**) Near-field PL spectra obtained in different positions under low-excitation conditions

To confirm whether these sharp PL peaks originate from the recombination of localized carriers (excitons), we measured the position dependence of the near-field PL spectra shown in Fig. 31b. These four PL spectra were obtained at different positions under low-excitation conditions. The number of sharp peaks and these energy positions in each spectrum are drastically different from position to position. These sharp emissions occur in a small area on a scale of within 150 nm, that is, in the local potential minimum, and the scale of the local potential fluctuation is estimated to be at least within 150 nm. Therefore, we conclude that the dominant emission mechanism at low temperature is recombination of the trapped and localized carriers (excitons) due to energy fluctuation. This energy fluctuation could come from the compositional variation of nitrogen, since the spatial distribution of the optical properties strongly depends on the nitrogen concentration and the scale of the inhomogeneity decreases with increase in nitrogen concentration.

8 Perspectives

The quality and reproducibility of NSOM measurements have been greatly improved in the last few years by the introduction of high-definition fiber probes made by combining chemical etching with ion-beam milling or by using a simpler impact method. Probes having apertures with diameters as small as 10 nm have been made, and their excellent performance has been demonstrated through high-resolution imaging of semiconductor nanostructures and single molecules. The probes, however, should be improved with respect to optical transmission, tip stability, and damage threshold. Microfabricated probes offer a promising solution to these difficulties and are likely to improve experimental reproducibility by providing better-defined conditions. Apertureless (or scattering) techniques will surely improve spatial resolution.

In this chapter we focused on the spectroscopic applications of NSOM and covered only a small portion of the research currently being pursued in the area of near-field optics. The utilization of NSOM techniques is not restricted to pure research; it has led to the development of a wide range of nano-optical devices. Real-space visualization of electronic quantum states will contribute to wavefunction engineering, where we deal with complicated potential structures for the design of novel devices. Insight into the nanometer-scale interaction between the mesoscopic electronic system and photons will provide us with the key concepts underlying the operation of nano-optoelectronic devices. For example, the near-field optical coupling of nano-electronic systems can enhance far-field forbidden transitions. Such alternation of the selection rule of optical transition will lead to spatiotemporal coherent control of electronic excitation, which can be utilized as the principal function of nanometric devices.

We are grateful to M. Ohtsu, S. Mononobe, K. Matsuda, N. Hosaka, H. Kambe, K. Sawada, H. Nakamura, T. Inoue, F. Sato, K. Nishi, H. Saito, Y. Aoyagi, M. Mihara, S. Nomura, M. Takahashi, A. Moto, and S. Takagishi for their assistance and fruitful discussions.

References

1. P. Borri, W. Langbein, S. Schneider, U. Woggon, R.L. Sellin, D. Ouyang, D. Bimberg: Phys. Rev. Lett. **87**, 157401 (2001)
2. D. Birkedal, K. Leosson, J.M. Hvam: Phys. Rev. Lett. **87**, 227401 (2001)
3. K. Brunner, G. Abstreiter, G. Bohm, G. Trankel, G. Weimann: Phys. Rev. Lett. **73**, 1138 (1994)
4. H.F. Hess, E. Betzig, T.D. Harris, L.N. Pfeiffer, K.W. West: Science **264**, 1740 (1994)
5. D. Gammon, E.S. Snow, B.V. Shanabrook, D.S. Katzer: Phys. Rev. Lett. **76**, 3005 (1996)
6. T.H. Stievater, X. Li, D.G. Steel, D. Gammon, D.S. Katzer, D. Park, C. Piermarocchi, L.J. Sham: Phys. Rev. Lett. **87**, 133603 (2001)
7. H. Kamada, H. Gotoh, J. Temmyo, T. Takagahara, H. Ando: Phys. Rev. Lett. **87**, 246401 (2001)
8. G. Chen, N.H. Bonadeo, D.G. Steel, D. Gammon, D.S. Katzer, D. Park, L.J. Sham: Science **289**, 1906 (2000)
9. P. Michler, A. Kiraz, C. Becher, W.V. Schoenfeld, P.M. Petroff, L. Zhang, E. Hu, A. Imamoglu: Science **290**, 2282 (2000)

10. M. Ohtsu: *Near-Field Nano/Atom Optics and Spectroscopy* (Springer-Verlag, Tokyo 1998)
11. M. Ohtsu: *Optical and Electronic Process of Nano-Matters* (Kluwer Academic Publishers, Tokyo 2001)
12. S. Kawata: *Near-Field Optics and Surface Plasmon Polaritons* (Springer-Verlag, Telos 2001)
13. V. Emiliani, T. Guenther, C. Lienau, R. Nötzel, K.H. Ploog: Phys. Rev. B **61**, R10583 (2000)
14. M. Achermann, B.A. Nechay, F. Morie-Genoud, A. Schertel, U. Siegner, U. Keller: Phys. Rev. B **60**, 2101 (1999)
15. N.H. Bonadeo, J. Erland, D. Gammon, D. Park, D.S. Katzer, D.G. Steel: Science **282**, 1473 (1998)
16. Y. Toda, T. Sugimoto, M. Nishioka, Y. Arakawa: Appl. Phys. Lett. **76**, 3887 (2000)
17. G.A. Valaskovic, M. Holton, G.H. Morrison: Appl. Opt. **34**, 1215 (1995)
18. T. Pangaribuan, K. Yamada, S. Jiang, H. Ohsawa, M. Ohtsu: Jpn. J. Appl. Phys. **31**, L1302 (1992)
19. S. Mononobe, M. Ohtsu: IEEE Photon. Technol. Lett. **10**, 99 (1998)
20. M. Muranishi, K. Sato, S. Hosaka, A. Kikukawa, T. Shintani, K. Ito: Jpn. J. Appl. Phys. **36**, L942 (1997)
21. T. Saiki, K. Matsuda: Appl. Phys. Lett. **74**, 2773 (1999)
22. T. Saiki, S. Mononobe, M. Ohtsu, N. Saito, J. Kusano: Appl. Phys. Lett. **68**, 2612 (1996)
23. H. Furukawa, S. Kawata: Opt. Commun. **132**, 170 (1996)
24. H. Nakamura, T. Sato, H. Kambe, K. Sawada, T. Saiki: J. Microscopy **202**, 50 (2001)
25. H.A. Bethe: Phys. Rev. **66**, 163 (1944)
26. K. Matsuda, T. Saiki, S. Nomura, M. Mihara, Y. Aoyagi: Appl. Phys. Lett. **81**, 2291 (2002)
27. D. Gammon, E.S. Snow, D.S. Katzer: Appl. Phys. Lett. **67**, 2391 (1995)
28. N. Hosaka, T. Saiki: J. Microsc. **202**, 362 (2001)
29. X.S. Xie, R.C. Dunn: Science **265**, 361 (1994)
30. T. Saiki, K. Nishi, M. Ohtsu: Jpn. J. Appl. Phys. **37**, 1639 (1998)
31. E. Dekel, D. Gershoni, E. Ehrenfreund, D. Spektor, J.M. Garcia, P.M. Petroff: Phys. Rev. Lett. **80**, 4991 (1998)
32. Y. Toda, O. Moriwaki, M. Nishioka, Y. Arakawa: Phys. Rev. Lett. **82**, 4114 (1999)
33. M. Notomi, T. Furuta, H. Kamada, J. Temmyo, T. Tamamura: Phys. Rev. B **53**, 15743 (1996)
34. K. Ota, N. Usami, Y. Shiraki: Physica E **2**, 573 (1998)
35. D. Gammon, E.S. Snow, B.V. Shanabrook, D.S. Katzer, D. Park: Science **273**, 87 (1996)
36. A.V. Uskov, K. Nishi, R. Lang: Appl. Phys. Lett. **74**, 3081 (1999)
37. H. Tsuchiya, T. Miyoshi: Solid-State Electron. **42**, 1443 (1998)
38. K. Matsuda, K. Ikeda, T. Saiki, H. Tsuchiya, H. Saito, K. Nishi: Phys. Rev. B **63**, 121304 (2001)
39. K. Nishi, R. Mirin, D. Leonard, G. Medeiros-Ribeiro, P.M. Petroff, A.C. Gossard: J. Appl. Phys. **80**, 3466 (1996)
40. H. Kamada, J. Temmyo, M. Notomi, T. Furuta, T. Tamamura: Jpn. J. Appl. Phys. **36**, 4194 (1998)

41. P. Borri, W. Langbein, J.M. Hvam, F. Heinrichsdorff, M.-H. Mao, D. Bimberg: Appl. Phys. Lett. **76**, 1380 (2000)
42. K. Matsuda, T. Saiki, H. Saito, K. Nishi: Appl. Phys. Lett. **76**, 73 (2000)
43. H. Tsuchiya, T. Miyoshi: Microelectron. Eng. **47**, 139 (1999)
44. D. Gammon, S. Rudin, T.L. Reinecke, D.S. Katzer, C.S. Kyono: Phys. Rev. B **51**, 16785 (1995)
45. H. Haug, S.W. Koch: *Quantum Theory of the Optical and Electronic Properties of Semiconductors* (World Scientific, Singapore 1993)
46. G. Trankel, E. Lach, A. Forchel, F. Scholz, C. Ell, H. Haug, G. Wimann: Phys. Rev. B **36**, 3712 (1987)
47. Q. Wu, R.D. Grober, D. Gammon, D.S. Katzer: Phys. Rev. B **62**, 13022 (2000)
48. R. Rinaldi, G. Coli, A. Passaseo, R. Cingolani: Phys. Rev. B **59**, 2230 (1999)
49. K. Nishi, H. Saito, S. Sugou, J.-S. Lee: Appl. Phys. Lett. **74**, 1111 (1999)
50. M. Sugawara, K. Mukai, Y. Nakata: Appl. Phys. Lett. **75**, 656 (1999)
51. R. Heitz, F. Guffarth, I. Mukhametzhanov, M. Grundmann, A. Madhukar, D. Bimberg: Phys. Rev. B. **62**, 16881 (2000)
52. S.V. Nair, Y. Masumoto: Phys. Stat. Sol. A **178**, 303 (2000)
53. S.V. Nair, Y. Masumoto: J. Lumin. **87–89**, 438 (2000)
54. A.V. Uskov, I. Magnusdottir, B. Tomborg, J. Mork, R. Lang: Appl. Phys. Lett. **79**, 1679 (2001)
55. R. Ferreira, G. Bastard: Appl. Phys. Lett. **74**, 2818 (1999)
56. J.R. Guest, T.H. Stievater, G. Chen, E.A. Tabak, B.G. Orr, D.G. Steel, D. Gammon, D.S. Katzer: Science **293**, 2224 (2001)
57. G.W. Bryant: Appl. Phys. Lett. **72**, 768 (1998)
58. G. Chen, T.H. Stievater, E.T. Batteh, X. Li, D.G. Steel, D. Gammon, D.S. Katzer, D. Park, L.J. Sham: Phys. Rev. Lett. **88**, 117901 (2002)
59. M. Weyers, M. Sato, H. Ando: Jpn. J. Appl. Phys. **31**, 853 (1992)
60. X. Yang, J.B. Heroux, M.J. Jurkovic, W.I. Wang: Appl. Phys. Lett. **76**, 795 (2000)
61. M. Kondow, S. Nakatsuka, T. Kitatani, Y. Yazawa, M. Okai: Jpn. J. Appl. Phys. **35**, 5711 (1996)
62. M. Kondow, K. Uomi, K. Hosomi, T. Mozume: Jpn. J. Appl. Phys. **33**, L1056 (1994)
63. M. Sato: J. Cryst. Growth **145**, 99 (1994)
64. I.A. Buyanova, W.M. Chen, G. Pozina, J.P. Bergman, B. Monemar, H.P. Xin, C.W. Tu: Appl. Phys. Lett. **75**, 501 (1999)
65. L. Grenouillet, C. Bru-Chevallier, G. Guillot, P. Gilet, P. Duvaut, C. Vannuffel, A. Million, A. Chenevas-Paule: Appl. Phys. Lett. **76**, 2241 (2000)
66. K. Uesugi, I. Suemune, T. Hasegawa, T. Akutagawa, T. Nakamura: Appl. Phys. Lett. **76**, 1285 (2000)
67. I.A. Buyanova, W.M. Chen, B. Monemar, H.P. Xin, C.W. Tu: Appl. Phys. Lett. **75**, 3781 (1999)
68. B.Q. Sun, D.S. Jiang, X.D. Luo, Z.Y. Xu, Z. Pan, L.H. Li, R.H. Wu: Appl. Phys. Lett. **76**, 2862 (2000)
69. R.A. Mair, J.Y. Lin, H.X. Jiang, E.D. Jones, A.A. Allerman, S.R. Kurtz: Appl. Phys. Lett. **76**, 188 (2000)
70. K. Matsuda, T. Saiki, M. Takahashi, A. Moto, S. Takagishi: Appl. Phys. Lett. **78**, 1508 (2001)
71. A. Moto, S. Tanaka, N. Ikoma, T. Tanabe, S. Takagishi, M. Takahashi, T. Katsuyama: Jpn. J. Appl. Phys. **38**, 1015 (1999)

72. M. Takahashi, A. Moto, S. Tanaka, T. Tanabe, S. Takagishi, K. Karatani, M. Nakayama, K. Matsuda, T. Saiki: J. Cryst. Growth **221**, 461 (2000)
73. T. Someya, Y. Arakawa: Jpn. J. Appl. Phys. **38**, L1216 (1999)

Atom Deflector and Detector with Near-Field Light

H. Ito, K. Totsuka, and M. Ohtsu

1 Introduction

Manipulating atoms with freedom has been one of any scientist's dreams for a long time. Atoms have been used as a measure of whether quantum mechanics is correct. As is well known, the frequency and wavelength of light emitted from a Cs atom or a Rb atom are adopted as standards of time and length. For these purposes, gaseous atoms are often used. However, such atoms move about actively with high speed in vacuum. The thermal motion at random formed a barrier against high-resolution laser spectroscopy. For example, some important signals are hidden away in Doppler-broadened spectra. Consequently, it was a critical issue to control the atomic motion.

The only way of controlling neutral atoms is to use interaction with light. When laser light is tuned to an atomic transition, it exerts resonant mechanical forces on atoms [1]. Indeed, such forces have been used for decelerating and cooling gaseous atoms [2–6]. The most famous technique is the magneto-optical trap (MOT) [7,8]. In MOT, atoms are decelerated by three orthogonal pairs of counterpropagating σ^+-σ^- circularly polarized light beams and trapped at the center where the six light beams cross by modulating the atomic Zeeman sublevels with anti-Helmholtz coils. The mean temperature of the cooled atoms drops to several μK when polarization gradient cooling (PGC) [9,10] is performed. The laser-cooling techniques have been applied to nonlinear spectroscopy [11], optical lattices [12], ultracold collisions [13–15], atomic fountains [16], atom interferometry [17,18], atom holography [19,20], quantum chaos [21], etc. One of the goals was realization of Bose–Einstein condensation (BEC) of alkali-metal atoms [22–24]. To this end, atoms were cooled further to the order of 10 nK by evaporative cooling and placed in a magnetic trap. Then, many theoretical and experimental studies on relevant topics were promoted. In particular, a coherent output of atomic waves from BEC has been observed, which is called an atom laser [23–25].

Mechanical forces of light on atoms are approximately divided into two parts. One is a spontaneous force induced by photon absorption followed by random scattering on resonance. For two-level atoms with an electric dipole moment μ, the spontaneous force $\boldsymbol{F}_{\mathrm{sp}}$ is given by [26]

$$\boldsymbol{F}_{\mathrm{sp}} = \hbar\gamma\boldsymbol{k}\frac{\Omega^2}{4\delta^2+\gamma^2+2\Omega^2}\,, \tag{1}$$

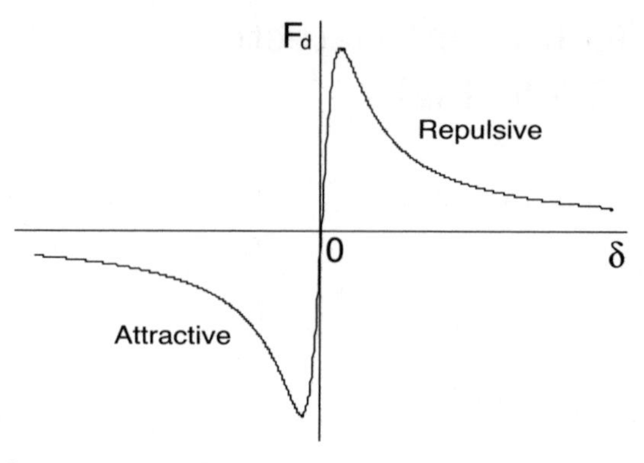

Fig. 1. Dipole force F_d plotted as a function of the frequency detuning δ. It is repulsive for $\delta > 0$ (blue detuning), while attractive for $\delta < 0$ (red detuning)

where $\hbar$ is the Planck constant, γ is the natural linewidth, $\boldsymbol{k}$ is the wavenumber vector, $\Omega = \mu E/\hbar$ with an electric-field amplitude E is the Rabi frequency, and $\delta = \omega_L - \omega_0$ is the detuning of the light frequency ω_L from the atomic resonance frequency ω_0. The spontaneous force works in the direction of light propagation, that is, in the direction of the wavenumber vector $\boldsymbol{k}$ and dissipates atomic kinetic energies. For an atom moving with a velocity $\boldsymbol{v}$, the frequency detuning $\Delta = \omega_L - \omega_0 - \boldsymbol{k} \cdot \boldsymbol{v}$ including a Doppler shift $\boldsymbol{k} \cdot \boldsymbol{v}$ is used. The spontaneous force depending on the atomic velocity is used for Doppler cooling in MOT [10]. The other is a dipole force, which is also called a gradient force. For two-level atoms, the dipole force $\boldsymbol{F}_d$ is given by [26]

$$\boldsymbol{F}_d = -\frac{\hbar\delta\nabla\Omega^2}{4\delta^2 + \gamma^2 + 2\Omega^2} \, . \tag{2}$$

As shown in Fig. 1, the dipole force has a dispersion character. When the light frequency is higher than an atomic resonance frequency, the dipole force works in the direction where the light intensity decreases. In the opposite case, it works in the direction where the light intensity increases. Namely, the dipole force is repulsive when $\delta > 0$ (blue detuning), while attractive when $\delta < 0$ (red detuning). The dipole force has been used as a mirror, a splitter, and a grating for atomic beams [27,28]. Moreover, guiding of atoms using hollow optical fibers has been proposed [29–32] and demonstrated [33–35]. In the atom guiding, blue-detuned evanescent light is induced around the inner-wall surface of the hollow core. Some kinds of atom guiding are introduced in [36,37]. A research field dealing with resonant interaction between light and atoms is called atom optics. Atom manipulation using laser light is discussed in [38].

Optical nanofabrication is a fascinating application of atom optics. Indeed, by focusing an atomic beam using atom-optical methods, small lines and arrays with each individual width of 10-nm order have been produced

[39–46]. Atom lithography has also been performed [47,48]. However, the conventional methods using light propagation have limited spatial accuracy of atom control due to the diffraction limit, which is about half the wavelength of the light used. The spread light field cannot move atoms to the exact position they are aimed at. It is also difficult to make arbitrary shapes using propagation-light methods. To our knowledge, dots have not been made using pure atom-optical methods.

In order to overcome the diffraction limit, one can use near-field light, which is not affected by the diffraction limit. Such near-field light is generated, for example, at the nanometric aperture of a sharpened fiber probe [49]. Since near-field light decays as a Yukawa-type function [50,51] and the dipole force is proportional to the spatial gradient of the light intensity, as shown in (2), near-field light exerts a strong dipole force on atoms. Using the dipole force from nanometric near-field light, one can precisely control atomic motion. We have proposed atom deflection using a fiber probe [52,53]. The feasibility of deflecting atoms with a fiber probe has been theoretically examined [54,55].

The atom-deflection technique can be applied to atom-by-atom deposition. Figure 2 schematically shows an optically controlled atom deposition on a substrate. The repulsive dipole force deflects atoms passing through blue-detuned near-field light induced at the tip of a fiber probe. Note that the deflection angle can be controlled by changing the blue detuning and the light intensity. Consequently, one can send atoms to the position on a substrate aimed at.

A small number of atoms contribute to the deflection using near-field light, so that it is difficult to detect the deflected atoms. In order to raise the detec-

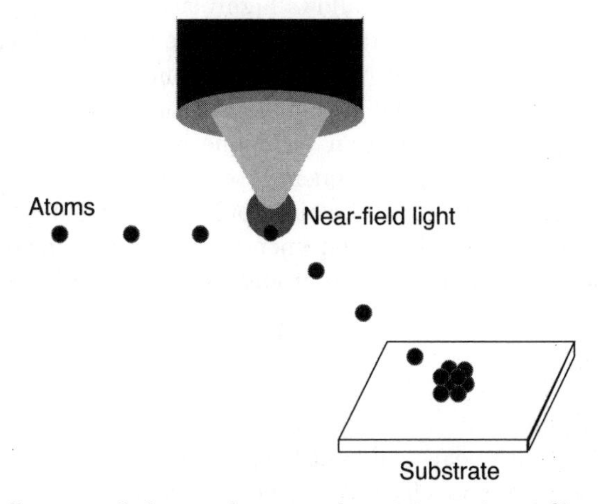

Fig. 2. Optically controlled atom-by-atom deposition using a fiber probe. Atoms are deflected by a repulsive dipole force from blue-detuned near-field light and sent to a point on a substrate that is aimed at. The deflection angle is controlled by the frequency detuning and the light intensity

tion efficiency, we developed a slit-type deflector [56]. Atoms are deflected by the repulsive dipole force from blue-detuned near-field light induced at the slit. The number of deflected atoms is increased by making a long slit. We also developed a new atom detector with high spatial resolution [57], which has a slit structure similar to the deflector. In the detector, two-color near-field lights are used for two-step photoionization or induction of blue fluorescence.

In Sect. 2, we introduce a slit-type atom deflector fabricated from a silicon-on-insulator substrate. The intensity distribution at the slit edge is measured using a fiber probe. From the intensity profile, we estimate the deflection angle of Rb atoms as a function of the frequency detuning and the light intensity. In Sect. 3, we introduce a slit-type atom detector. The intensity distribution is measured with a fiber probe again. A method of detecting atoms by two-step photoionization is first discussed and then the other method by observation of blue fluorescence is discussed. Next, we describe preliminary experiments of detecting Rb atoms by two-step photoionization and blue-fluorescence spectroscopy. From these experiments, we estimate the detection efficiencies for both schemes. High-temperature stability of the detector is also discussed. In Sect. 4, we describe experiments of guiding cold atoms to a slit-type deflector using a blue-detuned hollow light beam. In the scheme, Sisyphus cooling increases the guiding efficiency.

2 Slit-Type Deflector

In this section, we present a new scheme for deflecting atoms using blue-detuned near-field light generated at a long, narrow slit. Like the fiber-probe scheme, a slit-type deflector also allows highly accurate atomic spatial control. Moreover, it has the advantage of being able to deflect many atoms at once, so that it is easier to detect the deflected atoms than with a fiber probe. Consequently, it is useful for the first demonstration of precise atom control with near-field light. The deflection technique is used for optical nanofabrication. By adjusting the light frequency, one can selectively deposit atom species on a substrate [58]. It can be also applied to nondemolition measurement of optical near fields. This will deepen our understanding of microscopic and mesoscopic interactions between atoms and near-field light.

2.1 Principle

A pyramidal silicon probe has been developed for high-density, high-speed recording/reading with near-field light [59]. The pyramidal probe has a small aperture at the peak and efficiently generates nanometric near-field light with high throughput, which is defined as the conversion ratio from far-field light to near-field light. Considering this advantage, we fabricated a slit-type deflector from a pyramidal probe. Figure 3 illustrates the atom deflection using the slit-type deflector. A triangular-pillar structure is made on one side

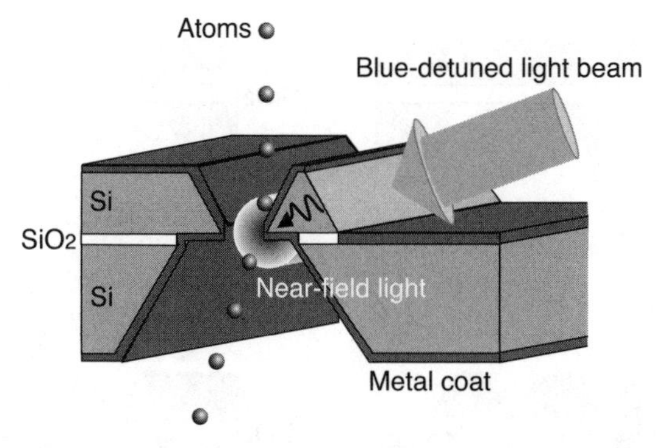

Fig. 3. Sketch of a slit-type atom deflector fabricated from a silicon-on insulator substrate. Only atoms entering the slit are deflected by the repulsive near-field light induced at an edge of the slit. Al-metal coating is made for suppression of far-field light

of the slit in order to introduce a light beam. Near-field light is generated at an edge of the slit by irradiating the backside surface with a light beam. Atoms passing through the slit are deflected by the repulsive dipole force under blue-detuning conditions. Atoms not entering the near-field light are blocked by the V-shaped groove in front of the slit, so that only deflected atoms leave the deflector. To suppress the generation of far-field light, Al metal is coated.

2.2 Fabrication Process

We made a slit-type deflector with a slit 100-nm wide and 100-µm long by photolithography and anisotropic chemical etching [60] of a (100)-oriented silicon-on insulator substrate. Figure 4 shows the process.

1. The central part of the lower side (the output side) is removed by photolithography and anisotropic etching with a 34-wt% KOH solution at a temperature of 80°C (Fig. 4a).
2. A V-shaped groove on the (111) face is formed on the upper side (the input side) by photolithography and KOH etching. The groove is 10 µm deep, with a slope length of 14 µm and an edge angle of 54.8°. Then, an incident surface for excitation light is made on the upper right-hand side by photolithography and KOH etching (Fig. 4b).
3. The SiO_2 surfaces are removed by etching with a buffered hydrofluoric acid. The residual Si layer on the lower side supports the slit structure (Fig. 4c).
4. An Al coating is formed by vacuum deposition of Al metal on both sides of the slit. It is used to suppress the generation of far-field light. As shown by arrows, the directional deposition is made to leave the light-incident surface and the top edge uncoated. In order to induce near-field light

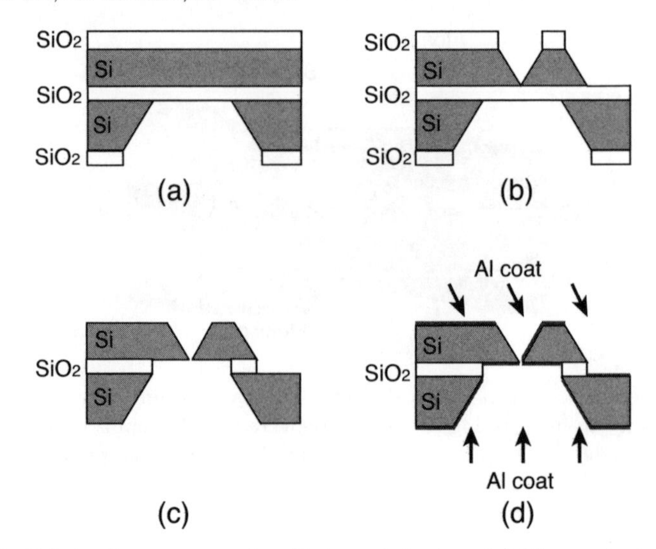

Fig. 4a-d. Fabrication process of a slit-type deflector. (**a**) The central part of the lower side is removed by photolithography and anisotropic etching. (**b**) A V-shaped groove is formed on the upper side, and an incident surface for excitation light is made on the upper right-hand side. (**c**) Extra SiO_2 layers are removed. (**d**) An Al coating is made from the direction indicated by *arrows*

with a localization length equal to the slit width, we adjusted the radius of curvature of the top edge to 50 nm with a coating thickness of 40 nm (Fig. 4d).

Since the slit width and length are easily determined by controlling the etching time, we can make smaller, shorter slits as needed. In this case, the slit width and length were determined so that the slit-type deflector could be combined with a slit-type atom detector mentioned in Sect. 3. Figure 5 shows a SEM image of the slit in the deflector. Its 100-nm width is comparable

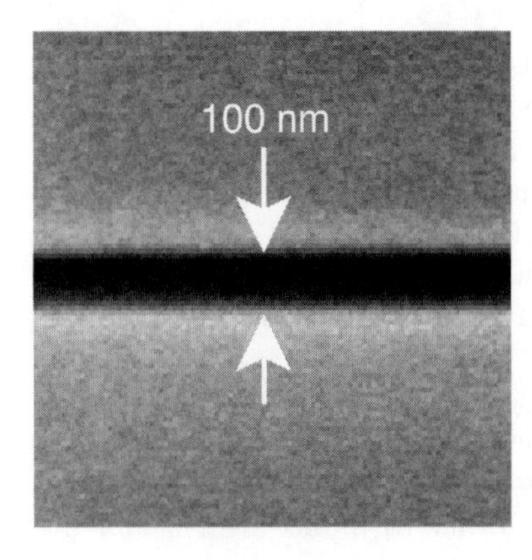

Fig. 5. SEM image of a 100-nm wide slit. The total length is 100 μm

to the spatial resolution of the detector. The deviation of a Rb atom from the incident axis is estimated to exceed 10 μm under realistic conditions [52]. This indicates that the deflected atoms can be detected with a spatial uncertainty of less than 1%. By contrast, the detection efficiency depends on the slit length. When the slit length of the detector is the same as that of the deflector, the detection efficiency is estimated to exceed 10% for laser-cooled atoms [57].

2.3 Measurement of Light Distribution

In order to examine the spatial distribution of the near-field light, we scanned the edge with a 100-nm aperture fiber probe. Figure 6 shows the configuration of the edge and the fiber probe. The origin of the x axis denoting the scanning direction was taken as 10 nm from the top edge. For simplicity, we assume that the uncoated slit edge is approximately hemicylindrical with radius a.

The solid curve A in Fig. 7 shows the light-intensity profile produced by a light beam with a wavelength of 780 nm (the transition wavelength of the Rb D_2 line), a power of 1 mW and a spot diameter of 10 μm. Here, the light beam is polarized perpendicular to the slit-length direction, since there is no cutoff and near-field light is generated efficiently [61]. In this measurement, we removed the opposite part of the slit where the near-field light is not induced to facilitate probe scanning. The distribution length of the near-field light component is 180 nm, as estimated below. However, note that this convolutes the size of the aperture. Incidentally, propagating far-field light appears in the outer region at a distance exceeding +90 nm. This is due to the imperfect metal coating of the bottom surface.

In order to evaluate the effective distribution length of the near-field light, we calculated the intensity profile using the phenomenological formula [50]

$$I_{\mathrm{nf}}(\boldsymbol{R}) = I_{\mathrm{nf}}(\boldsymbol{0})\frac{H(\boldsymbol{R})}{H(\boldsymbol{0})}\,, \tag{3}$$

where

$$H(\boldsymbol{R}) = |\nabla\Psi(\boldsymbol{R})|^2 + \frac{1}{\Lambda^2}|\Psi(\boldsymbol{R})|^2\,. \tag{4}$$

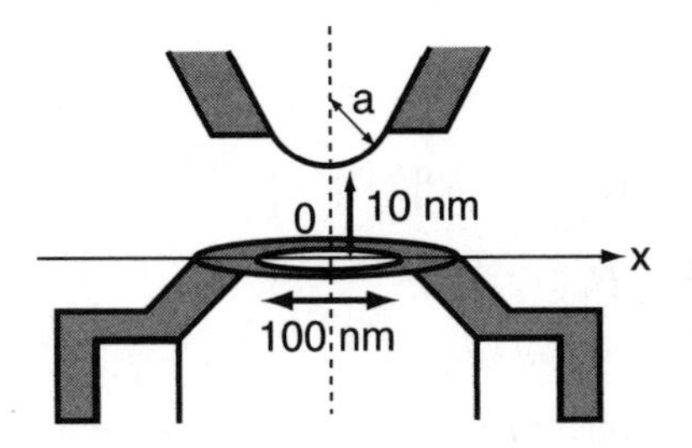

Fig. 6. Measurement configuration of the light-intensity distribution near the slit edge with a 100-nm aperture fiber probe. The scanning is performed along the x axis at 10 nm below the edge with radius a

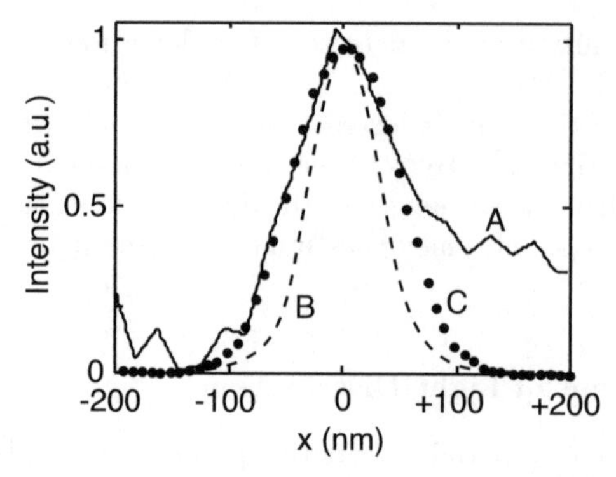

Fig. 7. The *solid curve A* shows the experimental result. The *broken curve B* shows the numerical result when the aperture size is not convoluted. The *dotted curve C* shows the numerical result when the aperture size is convoluted. In case A, a far-field component arises in the region exceeding +90 nm. Here, a light beam with a wavelength of 780 nm is used

A Yukawa-type function $\Psi(\boldsymbol{R})$ is given by

$$\Psi(\boldsymbol{R}) = \int \frac{\exp\left(-|\boldsymbol{R}-\boldsymbol{R}'|/\Lambda\right)}{|\boldsymbol{R}-\boldsymbol{R}'|} \mathrm{d}S \, . \tag{5}$$

The coordinate vectors $\boldsymbol{R}$ and $\boldsymbol{R}'$ indicate the measurement point on the aperture of the probe and the source point on the slit edge, respectively. The origin is at the top edge. The surface integral is made over the area $\pi a \times L$, where L is the excitation length equal to the spot diameter of the excitation light beam. For a decay length $\Lambda = a = 50$ nm, when we calculate $I_{\mathrm{nf}}(\boldsymbol{R})$, we obtain the broken curve B in Fig. 7. From the numerical result, the effective distribution length is estimated to be 126 nm, which is defined as the full width at the e^{-2} maximum. Moreover, integrating

$$I_{\mathrm{nf}} = \int I_{\mathrm{nf}}(\boldsymbol{R}) \mathrm{d}S \tag{6}$$

over the aperture area $\pi \times 50^2$ nm^2, we obtain the dotted curve C in Fig. 7. This curve closely matches the experimental curve A. The full width at the e^{-2} maximum of the profile C is 180 nm.

2.4 Estimation of Deflection Angle

The deflection angle θ of a ballistic atom is given by [62]

$$\theta = \pi - 2b \int_{r_{\mathrm{t}}}^{\infty} \frac{\mathrm{d}r}{r^2} \left(1 - \frac{b^2}{r^2} - \frac{U_{\mathrm{tot}}(r)}{K_{\mathrm{a}}}\right)^{-1/2} , \tag{7}$$

where b, r_{t}, and K_{a} are the impact parameter, turning point, and atomic kinetic energy, respectively. The distance r is measured from the center of the hemicylindrical slit-edge. The total potential $U_{\text{tot}}(r)$ consists of the repulsive dipole-force potential $U_{\text{dip}}(r)$ given by [26,36]

$$U_{\text{dip}}(r) = \frac{1}{2}\hbar\delta \ln\left(1 + \frac{I(r)}{I_{\text{s}}}\frac{\gamma^2}{4\delta^2 + \gamma^2}\right) \tag{8}$$

and the attractive van der Waals potential $U_{\text{vdw}}(r)$ given by [63–65]

$$U_{\text{vdw}}(r) = -\frac{1}{16(r-a)^3}\sum_j \frac{\hbar\gamma_j}{k_j^3}\frac{n_j^2 - 1}{n_j^2 + 1} \,. \tag{9}$$

The natural linewidth γ and the saturation intensity I_{s} are $2\pi \times 6.1$ MHz and 1.6 mW/cm^2 for the Rb D_2 line, respectively. The van der Waals potential is summed over the allowed dipole transitions labeled with j. Each natural linewidth and wavenumber are γ_j and k_j, respectively. The refractive index n_j of the slit edge is 3.7 for the Rb D_2 line.

Figure 8 shows the maximum deflection angle $\theta_{\max}$ plotted as a function of the atomic velocity v and the near-field light intensity $I_{\text{nf}}(\mathbf{0})$, where the atomic velocity is log-plotted. For simplicity, instead of the exact Yukawa-type function, we used the approximation

$$I_{\text{nf}}(r) = I_{\text{nf}}(\mathbf{0}) \exp\left(-\frac{r-a}{1.6 \times 10^{-8}}\right) \tag{10}$$

with $\delta = +2\pi \times 1$ GHz. The approximation is obtained from the curve B in Fig. 7. Since the throughput at the triangular pillar is about 0.01, as the excitation-light power changes from 1 mW to 10 mW, the near-field light intensity changes from 0.1 kW/cm^2 to 1 kW/cm^2. Note that slow atoms with a velocity of less than 10 m/s are easily generated with MOT. Such cold atoms are required for efficient interaction with nanometric near-field light. The maximum deflection angle $\theta_{\max}$ increases as the intensity $I_{\text{nf}}(\mathbf{0})$ increases and also as the atomic velocity v decreases. When $I_{\text{nf}}(\mathbf{0}) = 0.5$ kW/cm^2, a Rb atom with $v = 1$ m/s is retroreflected.

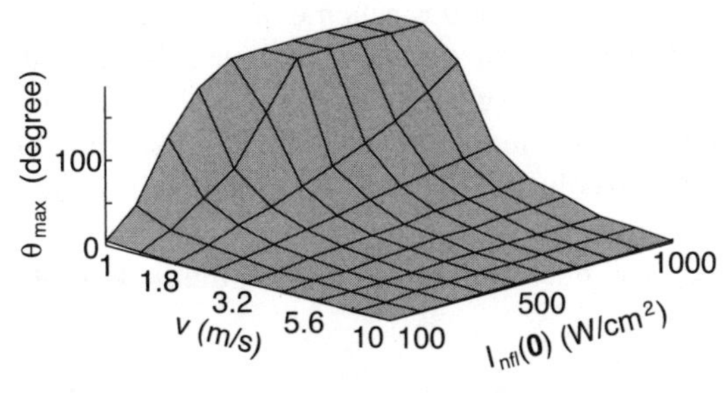

Fig. 8. Maximum deflection angle $\theta_{\max}$ of a Rb atom plotted as a function of the atomic velocity v and the near-field-light intensity $I_{\text{nf}}(\mathbf{0})$ under a frequency detuning $\delta/2\pi = +1$ GHz. The atomic velocity is log-plotted

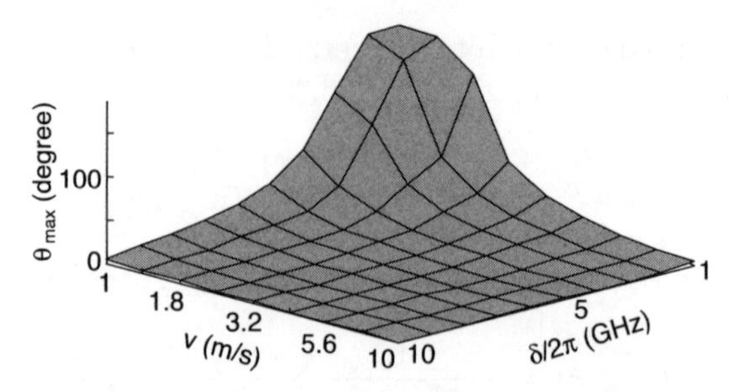

Fig. 9. Maximum deflection angle $\theta_{\max}$ of a Rb atom plotted as a function of the atomic velocity v and the frequency detuning $\delta/2\pi$ for $I_{\rm nf}(\mathbf{0}) = 10^3$ W/cm^2. The atomic velocity is log-plotted

Figure 9 shows the maximum deflection angle $\theta_{\max}$ plotted as a function of the atomic velocity v and the frequency detuning $\delta/2\pi$ for $I_{\rm nf}(\mathbf{0}) = 1$ kW/cm^2. When the frequency detuning changes from +10 GHz to +1 GHz, the maximum deflection angle $\theta_{\max}$ changes from 5.3° to 180° when $v = 1$ m/s.

3 Slit-Type Detector

A popular method of detecting neutral atoms is to use a microchannel plate (MCP). However, the highest resolution of commercial MCPs is only 5 μm. Although an elaborate detection system including a secondary-electron multiplier with a resolution of 1 μm has been reported [66], it is effective for metastable atoms but not for the ground-state atoms we manipulate. These detectors use surface ionization and the ionization efficiency for the ground-state atoms is estimated to be at most 10^{-4} per second at room temperature from the Saha–Langmuir equation and the adsorption time [67,68]. Thus, the conventional atom detectors have low detection efficiency as well as low spatial resolution, so that they are insufficient for the atom-deflection experiments using nanometric near-field light.

In this section, we describe a new scheme of detecting ground-state atoms, where a slit-type detector with a 70-nm wide slit is illuminated with two-wavelength laser beams for the generation of two-color near-field lights. Atoms are selectively detected by interaction with the two-color near-field lights. Note that the spatial resolution is principally determined by the localization length of near-field light, which approximately equals the slit width. Besides, in order to increase the detection efficiency, we made a 100-μm long slit.

3.1 Principle

Like the slit-type deflector, we fabricated a slit-type detector from a pyramidal probe. Figure 10 illustrates the atom detector made up of a Si layer

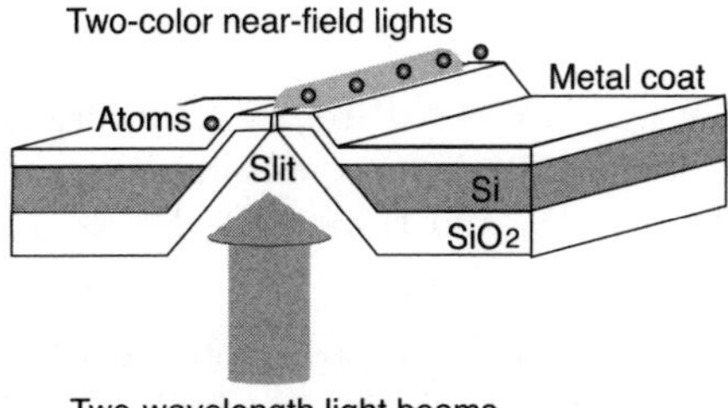

Fig. 10. Atom detection using a slit-type detector fabricated from a silicon-on-insulator substrate. Two-color near-field lights are induced at a nanometer wide slit by illuminating a V-shaped groove with two-wavelength light beams. Only atoms entering the near-field region are excited and detected. Here, the light beams are polarized perpendicular to the slit-length direction for high throughput

and a SiO_2 layer. Contrary to the Si layer, the SiO_2 layer hardly absorbs excitation light beams because the absorption coefficient is very small for the wavelengths used. To suppress the generation of far-field light, Al metal is coated.

In order to detect ground-state atoms efficiently and selectively, we use two-color near-field lights. When two-wavelength laser beams are introduced into a V-shaped groove, two-color near-field lights are induced at the nanometric slit. When the laser beams are polarized perpendicular to the slit-length direction, there is no cutoff due to the V-shaped groove and near-field light is generated efficiently [61]. Atoms approaching the slit interact with the two-color near-field lights and are detected selectively. In this case, the nanometric slit width ensures the high spatial resolution. By contrast, the slit length is long, so that the interaction time and opportunity between atoms and near-field light are increased.

For detection of Rb atoms, we use two methods. One is two-step photoionization [57]. The Rb atoms in the $5S_{1/2}$ upper ground state of the hyperfine structure are first excited to the $5P_{3/2}$ state and then to the ionization level by two-color near-field lights. The ionized atoms are counted using a channel electron multiplier (CEM) negatively biased. The other is stepwise resonant excitation of the $5S_{1/2}$ ground-state Rb atoms to the $5D_{5/2}$ state via the $5P_{3/2}$ state [69]. The Rb atoms in the $5D_{5/2}$ state can decay into the $5S_{1/2}$ ground state via the $6P_{3/2}$ state. In this case, the Rb atoms emit blue fluorescence with a wavelength of 420 nm. By monitoring the fluorescence, one can efficiently detect Rb atoms. The detection efficiency of cold Rb atoms is estimated to exceed 0.25 for the two-step photoionization and 0.03 for the fluorescence observation, as mentioned below. These detection efficiencies are sufficient for the atom-deflection experiments.

3.2 Fabrication Process

The slit-type detector was fabricated from a (100)-oriented silicon-on-insulator substrate by photolithography and anisotropic etching. The silicon-on-insulator substrate has two Si and three SiO_2 layers. Figure 11 shows the fabrication process.

1. First, a V-shaped groove is formed by photolithography and anisotropic etching with a 34-wt% KOH solution at 80°C (Fig. 11a).
2. Secondly, the top SiO_2 layer is cleared by selective etching with a buffered hydrofluoric (BHF) acid. Thirdly, the naked Si-layer surface is thermally oxidized with wet oxygen at 1150°C, so that a new SiO_2 layer covering the V-shaped groove is made on the Si layer (Fig. 11b).
3. Fourthly, a Cr coating was made on the SiO_2 layer by vacuum deposition of Cr metal. Fifthly, part of the SiO_2 layer located outside of the V-shaped groove is eliminated. Sixthly, a Pyrex glass plate is bonded to the open part of the Si layer by anodic bonding with 300 V at 350 °C. The Cr coating protects the SiO_2 layer on the V-shaped groove from percolation of a BHF acid which is used at (d) for perforating the slit. The Pyrex glass plate protects the fragile thin slit structure (Fig. 11c).
4. Seventhly, the extra SiO_2 and Si layers used as support are removed. Eighthly, the residual Si layer is etched with tetramethyl ammonium hydroxide until the peak of the V-shaped groove appears. Ninthly, the slit

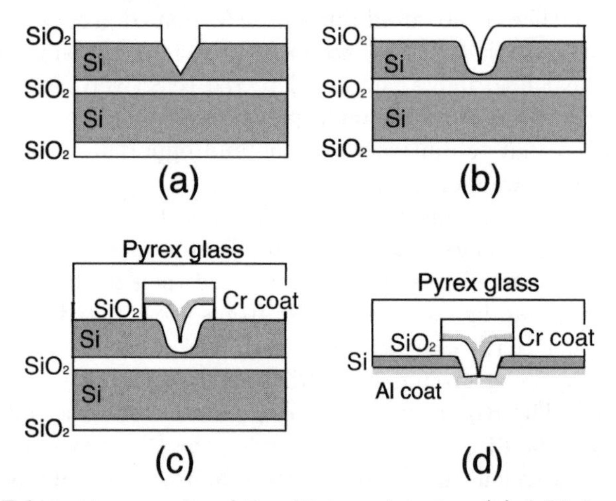

Fig. 11a-c. Fabrication process of the slit-type detector. (**a**) A V-shaped groove is formed by photolithography and anisotropic etching. (**b**) The top SiO_2 layer is first removed and then a new SiO_2 layer covering the V-shaped groove is made on the Si surface. (**c**) The SiO_2 layer is coated with Cr and then part outside the V-shaped groove is eliminated. Next, a Pyrex glass plate is bonded. (**d**) After removal of extra SiO_2 and Si layers, the residual Si layer is etched until the peak of the V-shaped groove appears. Then, the slit is made at the peak and the underside of the Si layer is coated with Al

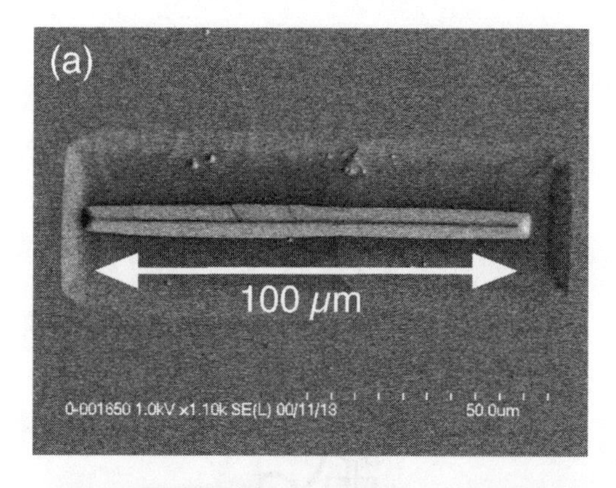

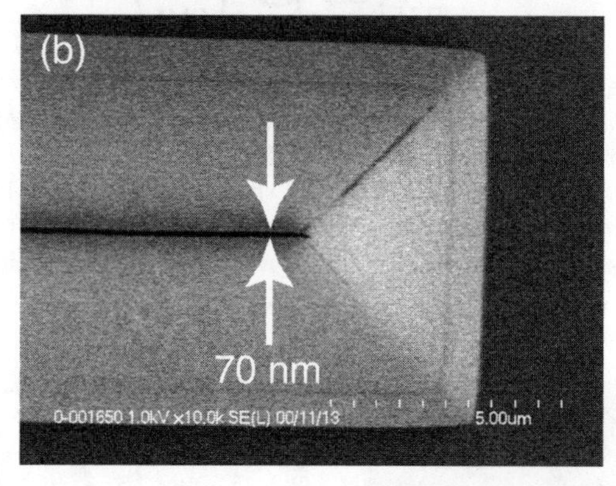

Fig. 12a,b. SEM images of a slit-type detector. The *upper* picture (**a**) shows a 100-μm long slit. As shown in the *lower* picture (**b**), the width is 70 nm

is made at the peak by slight etching with a BHF acid. Finally, the underside of the Si layer is coated with Al of 50-nm thickness (Fig. 11d).

Figure 12 shows the SEM images of a slit-type detector. The slit width is 70 nm, while the slit length is 100 μm for enhancement of the detection efficiency. The slit length can be controlled between 100 nm and 100 μm with the dispersion of about ±30 nm by changing the length of the mask for photolithography. The height of the V-shaped groove is 10 μm and the foot width is 14 μm.

3.3 Measurement of Light Distribution

The intensity distribution of near-field light generated at the slit was measured using a fiber probe. Figure 13 shows the experimental configuration. A 780-nm diode laser (LD) beam is coupled to a slit-type detector and near-field light is induced at the peak of the slit. A fiber probe with a small aperture of 70 nm is mounted on a tuning fork and controlled with a piezoelectric

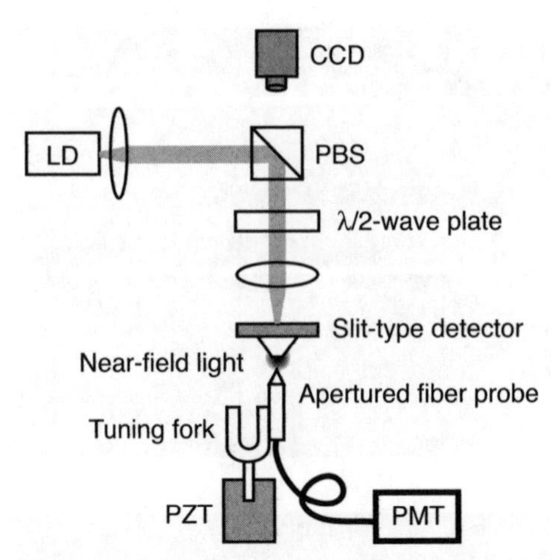

Fig. 13. Experimental setup for measurement of a light-intensity distribution. An apertured fiber probe is position controlled by the shear-force technique using a tuning fork and a piezoelectric transducer (PZT). The fiber probe collects scattered photons from near-field light generated at the slit. The diode laser (LD) beam introduced from the side using a polarization beam splitter (PBS) is polarized perpendicular to the slit-length direction through a $\lambda/2$-wave plate and coupled to the slit-type detector. A CCD camera monitors alignment of the light beam

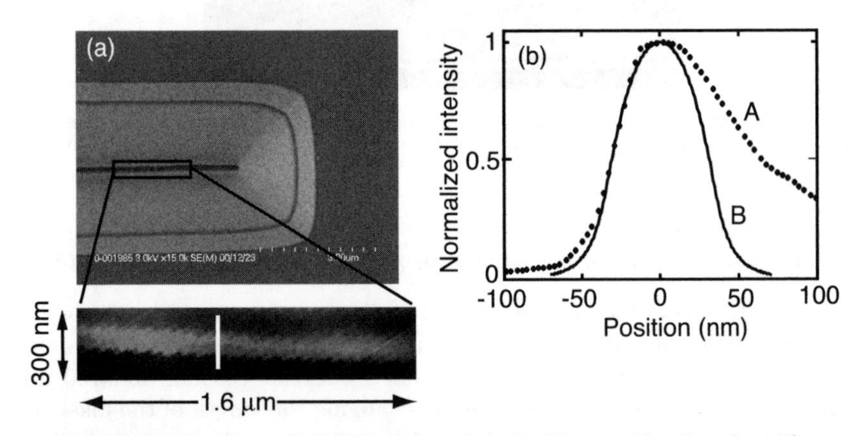

Fig. 14. (**a**) Image of near-field light induced at the 70-nm wide slit, where the area of 300 nm × 1.6 µm is scanned. (**b**) Intensity profile along the *white solid line* drawn in (**a**), where the intensity is normalized to the peak value. The *dotted curve A* with an FWHM of 96 nm shows the experimental result, while the *solid curve B* with an FWHM of 65 nm shows the theoretical profile

transducer (PZT). When the fiber probe goes into the near-field region, far-field photons are scattered and collected by the fiber probe. The output signal is sent to a photomultiplier tube (PMT) connected to the fiber probe. Here, the shear-force control monitoring the resonance frequency of the fiber-probe dithering is used for measurement of the distance d between the probe tip and the slit.

Figure 14a shows a SEM image of a slit-type detector and an image of near-field light generated at the slit. The latter is obtained from a 300-nm$\times$1.6-μm-size raster scanning at $d = 5$ nm. The dotted curve A in Fig. 14b shows the intensity profile in the direction designated by the white solid line. The FWHM of the curve A is estimated to be 96 nm.

We calculated the intensity profile using (3)–(5). The integration of (5) is carried out over the slit area $a_{\text{slit}} \times L_{\text{slit}}$, where a_{slit} and L_{slit} are the width and length of the slit, respectively. The solid curve B in Fig. 14b shows the intensity distribution at $d = 5$ nm for $\Lambda = a_{\text{slit}}/2 = 35$ nm and $L_{\text{slit}} = 100$ μm. The numerical result closely matches the experimental result on the left-hand side. By contrast, there is a deviation on the right-hand side. This is leakage of the light passing through the slit from the back, due to the imperfect Al coating. The FWHM of the curve B is 65 nm, which almost equals the slit width and gives the theoretical spatial resolution.

3.4 Two-Step Photoionization with Two-Color Near-Field Lights

Photoionization is one of the most effective methods of observing neutral atoms [70]. Here, we introduce two-step photoionization using two-color near-field lights. This scheme allows species-selective detection with high spatial resolution.

Method

As shown in Fig. 10, when the nanometric slit is illuminated with two kinds of laser beam, two-color near-field lights are induced at the slit. For Rb atoms, a 780-nm diode laser (LD) beam and a 476.5-nm Ar-ion laser beam are used for the generation of two-color near-field lights. Figure 15 shows the relevant energy levels of ^{85}Rb. The $5S_{1/2}$ ground state splits into two hyperfine levels labeled by quantum numbers $F = 2$ and 3. Similarly, the $5P_{3/2}$ excited state splits into four hyperfine levels with $F = 1, 2, 3$, and 4. The first near-field light generated by LD tuned to the $5S_{1/2}$, $F = 3 \rightarrow 5P_{3/2}$, $F = 4$ transition excites the ^{85}Rb atoms in the $5S_{1/2}$, $F = 3$ upper ground state to the $5P_{3/2}$, $F = 4$ state. Then, the second near-field light generated by the Ar-ion laser lifts the ^{85}Rb atoms in the $5P_{3/2}$, $F = 4$ state to the ionization level at 4.18 eV above the $5S_{1/2}$ ground state. The ionized atoms with positive charge are captured by a channel electron multiplier (CEM) with a negative bias of -3 kV (see Fig. 16). Since this scheme uses the resonant excitation, one can detect only Rb atoms state-selectively.

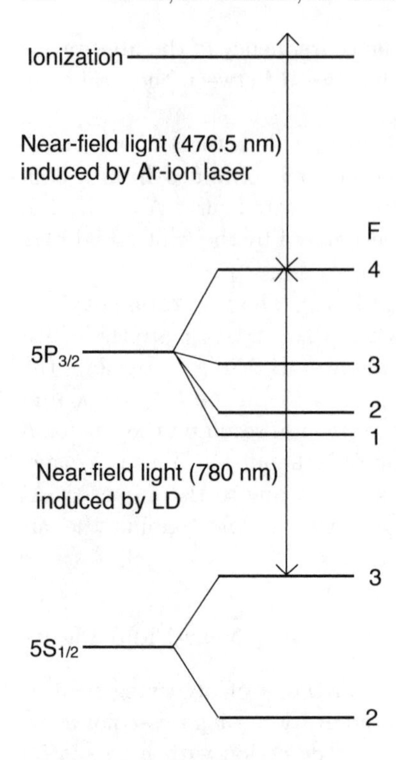

Fig. 15. Relevant hyperfine energy levels of ^{85}Rb. The 1st near-field light generated by a 780-nm diode laser (LD) induces the $5S_{1/2}$, $F = 3 \rightarrow 5P_{3/2}$, $F = 4$ transition. The 2nd near-field light generated by a 476.5-nm Ar-ion laser induces excitation to the ionization level

Experiment with Evanescent Light

In order to examine the detection efficiency, we conducted a two-step photoionization experiment of a Rb atomic beam. Figure 16 shows the experimental configuration. A planar near-field light (evanescent light) is generated at a prism surface via total internal reflection of an Ar-ion laser beam with

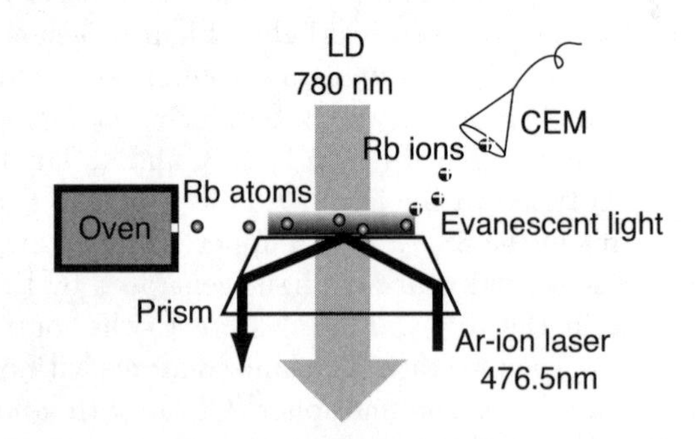

Fig. 16. Experimental configuration for two-step photoionization of a Rb atomic beam. Evanescent light is induced at a prism surface via total internal reflection of a 476.5-nm Ar-ion laser beam. A thermal Rb atomic beam from a 120 °C oven is ionized when a 780-nm diode laser (LD) beam is incident. The ionized Rb atoms are counted by a channel electron multiplier (CEM) with a negative bias of −3 kV

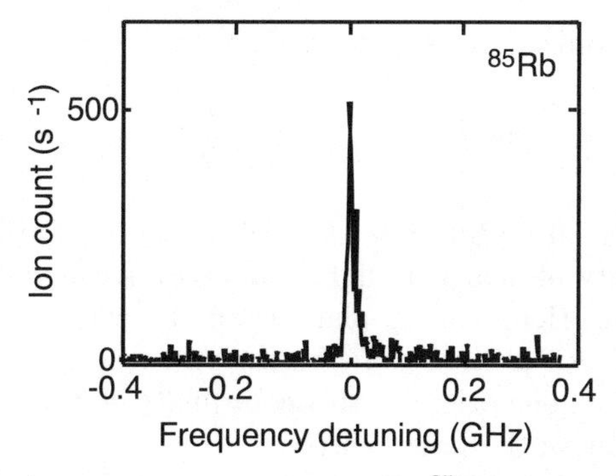

Fig. 17. Two-step photoionization spectrum for ^{85}Rb in the $5S_{1/2}$, $F = 3$ upper ground state. The ion number counted per second is plotted as a function of the LD frequency detuning, which is measured with respect to the $5S_{1/2}$, $F = 3 \rightarrow 5P_{3/2}$, $F = 4$ transition

a peak intensity of 2.0 kW/cm^2 under a vacuum pressure of 10^{-9} torr. Here, the light beam is polarized parallel to the plane of incidence. The penetration depth of the evanescent light is given by [36,71,72]

$$L_{\mathrm{pd}} = \frac{\lambda}{2\pi\sqrt{n^2 \sin^2\theta - 1}} , \tag{11}$$

where λ, θ, and n are the wavelength, the incident angle, and the refractive index of the prism, respectively. For $\lambda = 476.5$ nm, $\theta = 46°$, and $n = 1.52$, we obtain $L_{\mathrm{pd}} = 193$ nm. An LD beam with a peak intensity of 9 W/cm^2 and a beam spot w_{LD} of 100 μm is incident perpendicular to the prism surface. A Rb atomic beam with a flux intensity of 7.4×10^{14} m^{-2} s^{-1} and a diameter of 300 μm is generated by an oven with a temperature of 120 °C. When the atoms pass by the prism surface, they are ionized and attracted by a CEM biased at -3 kV. The quantum efficiency of the CEM is 0.9 for this bias.

Figure 17 shows a photoionization spectrum for the ^{85}Rb atoms in the $5S_{1/2}$, $F = 3$ upper ground state. The peak value of the ion count is about 500 per second on resonance. In this case, if we consider that the number of atoms passing through the interaction area $L_{\mathrm{pd}} \times w_{\mathrm{LD}}$ is 1.5×10^4 per second, we obtain a detection efficiency of 0.03.

Estimation of Detection Efficiency

The detection efficiency of the slit-type detector can be estimated from the experimental result mentioned above. To this end, we first evaluate the cross section σ_{ion} of the near-field light ionization from the $5P_{3/2}$ excited state. Since the LD intensity saturates the $5S_{1/2} \rightarrow 5P_{3/2}$ transition, we detect half of the Rb atoms in the $5P_{3/2}$ state on average. Consequently, the number

N_{ion} of the atoms ionized per second is given by

$$N_{\text{ion}} = \int_{-\infty}^{\infty} \mathrm{d}x \int_{-\infty}^{\infty} \mathrm{d}y \int_{0}^{\infty} \mathrm{d}z \int_{0}^{\infty} \mathrm{d}v \frac{N(y,z)}{2} \sigma_{\text{ion}} \phi(x,y,z) \frac{f(v)}{v} , \tag{12}$$

where $N(y,z)$ is the atom flux intensity, $\phi(x,y,z)$ is regarded as the virtual photon flux intensity of near-field light, and $f(v)$ is the velocity distribution of the atomic beam. Here, the xy-plane is on the prism surface. The x axis equals the incident axis of the atomic beam, while the z axis is perpendicular to the prism surface. The origin of this coordinate system is at the center of the near-field region on the prism surface.

Since the penetration depth $L_{\text{pd}} = 193$ nm is sufficiently small compared to the atomic beam diameter $l_{\text{atom}} = 300$ μm, the atom flux intensity is independent on z and expressed using a Gaussian function as

$$N(y,z) = N_0 \exp\left(-\frac{y^2}{l_{\text{atom}}^2}\right) , \tag{13}$$

where $N_0 = 7.4 \times 10^{14}$ m^{-2} s^{-1}. On the other hand, the virtual photon flux intensity is given by

$$\phi(x,y,z) = \phi_0 \exp\left(-\frac{2x^2}{w_x^2} - \frac{2y^2}{w_y^2} - \frac{2z}{L_{\text{pd}}}\right) , \tag{14}$$

where ϕ_0 is the value at the origin, and $w_x = 300$ μm and $w_y = 200$ μm are the major and minor axes of the elliptical Ar-ion laser beam, respectively. From the Fresnel laws on transmission, the conversion equation from far-field light to near-field light is derived as [73]

$$I_0 = \frac{4n\cos^2\theta}{(n^2-1)\left[(n^2+1)\sin^2\theta - 1\right]^p} I_{\text{i}} , \tag{15}$$

where I_{i} is the intensity of the Ar-ion laser light, $p = 0$ and 1 for the TE and TM polarizations, respectively. From (15), if the light polarization is parallel to the plane of incidence ($p = 1$), it follows that $\phi_0 = 3.1\phi_{\text{i}}$, where $\phi_{\text{i}} = 2.4 \times 10^{26}$ m^{-2} s^{-1} is the photon flux intensity of the Ar-ion laser light. The velocity distribution $f(v)$ is given by

$$f(v) = 2\left(\frac{m}{2k_{\text{B}}T}\right)^2 \exp\left(-\frac{mv^2}{2k_{\text{B}}T}\right) v^3 , \tag{16}$$

where v, m, T, and k_{B} are the atomic velocity, atomic mass, oven temperature, and Boltzmann constant, respectively. Substituting the experimental value of $N_{\text{ion}} \simeq 500$ s^{-1} and (13)–(15) into (16), we obtain $\sigma_{\text{ion}} = 2.4 \times 10^{-18}$ cm^2.

Now, let us consider the case of detecting slow ^{85}Rb atoms with the slit-type detector. If the intensity of near-field light generated by the LD

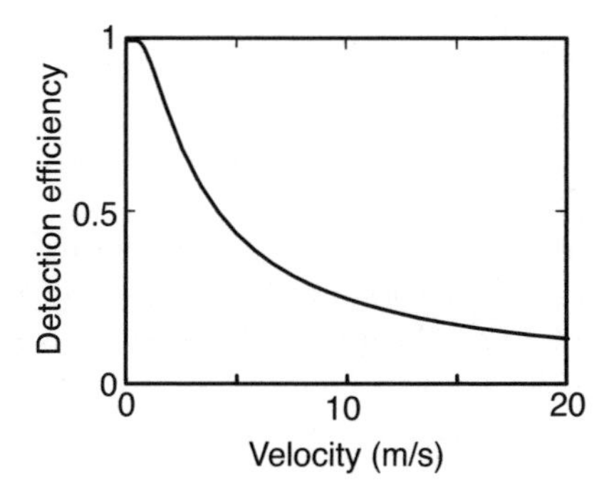

Fig. 18. Detection efficiency of the slit-type detector for slow Rb atoms plotted as a function of the atomic velocity

beam saturates the $5S_{1/2} \rightarrow 5P_{3/2}$ transition, the detection efficiency $\eta(v)$ is given by

$$\eta(v) = 1 - \exp\left(-\frac{\sigma_{\rm ion} I_{\rm nf} L_{\rm slit}}{2v}\right) , \tag{17}$$

where $I_{\rm nf}$ is the intensity of near-field light generated by the Ar-ion laser beam. Figure 18 shows the detection efficiency plotted as a function of the atomic velocity, where we approximate $I_{\rm nf} \simeq 2.4 \times 10^{26}\ {\rm cm^{-2}\,s^{-1}}$. For example, the detection efficiency at $v = 10$ m/s is estimated to be 0.25.

High-Temperature Stability

The ionization cross section $\sigma_{\rm ion}$ is small for the Rb atom in the $5P_{3/2}$ state. Consequently, the two-step photoionization requires a high-power Ar-ion laser beam exceeding 1 W. By making the beam waist smaller, one can increase the detection efficiency. However, such high-intensity light damages the slit-type detector by heating.

Let us estimate the critical intensity at which the Al coating melts. The Ar-ion laser beam is reflected by the Al coating with a reflectance $R = 0.9$ at the V-shaped groove. The loss is transformed into a thermal energy. Comparing the thermal conduction $\kappa\tau$ of each layer at the V-shaped groove, where the thermal conductivity $\kappa = 138$ W/cm K and the layer thickness $\tau = 4$ µm for the Si layer, $\kappa = 1$ W/cm K and $\tau = 1$ µm for the SiO_2 layer, and $\kappa = 240$ W/cm K and $\tau = 50$ nm for the Al coating, we find that $(\kappa\tau)_{\rm Si}$ is 46 times larger than $(\kappa\tau)_{\rm Al}$ and $(\kappa\tau)_{\rm Al}$ is 12 times larger than $(\kappa\tau)_{\rm SiO_2}$. Therefore, we consider the transmission of a thermal energy from the Al coating to the Si layer, neglecting that to the SiO_2 layer.

The thermal energy flux intensity J between two points with the distance Δs and the temperature difference ΔT is given by

$$J = \kappa \frac{\Delta T}{\Delta s} . \tag{18}$$

For simplicity, we assume that the thermal energy inflows from the window $L_{\rm slit} \times s$, where s is the slope length of the Al coating contacting with the

SiO_2 layer at the V-shaped groove. From (18), we obtain

$$I_{\mathrm{i}}(1-R)L_{\mathrm{slit}}s = \kappa \frac{\Delta T}{\Delta s} L_{\mathrm{slit}}\tau \, . \tag{19}$$

Now, let us consider the case where the temperature of the Al coating reaches the melting point of 660°C. If the Si layer connects to the body of a vacuum chamber at a room temperature of 30°C, the critical intensity $I_{\mathrm{i}}^{\mathrm{c}}$ is estimated to be 150 kW/cm^2 from (19) with $\Delta T = 630$ K and $s = \Delta s = 7$ μm. In the critical case, the detection efficiency is 0.98 for $v = 10$ m/s.

3.5 Blue-Fluorescence Spectroscopy with Two-Color Near-Field Lights

The slit-type detector gains a high detection efficiency for ground-state atoms by using two-step photoionization. Thanks to the resonant character, one can detect a specific atom state-selectively. However, a high-power laser exceeding 1 W is required for ionization from a low-lying energy level. Such an intense laser beam may thermally damage the slit-type detector as discussed above. Besides, use of an ion detector like CEM complicates the vacuum-chamber system.

Here, we describe a different scheme for detection of a small number of atoms using the slit-type detector. The alternative method requires no intense laser beam. In this case, Rb atoms are detected from fluorescence of the 2nd D_2 line induced by two-color near-field lights, which are generated by two low-power diode lasers with different wavelengths. By monitoring the blue fluorescence different from the infrared excitation laser light, one can detect Rb atoms without being disturbed by scattered light.

Method

Figure 19 shows the relevant energy levels of Rb for blue-fluorescence excitation. The first near-field light generated by a 780-nm diode laser (LD1) beam stimulates the $5S_{1/2} \to 5P_{3/2}$ transition. Then, the second near-field light generated by a 775-nm diode laser (LD2) beam stimulates the $5P_{3/2} \to 5D_{5/2}$ transition. The excited Rb atoms spontaneously decay to the $6P_{3/2}$ state or the $5P_{3/2}$ state with a branching ratio of 3 to 5 in a lifetime of 1.56 μs. The Rb atoms in the $6P_{3/2}$ state decay again to one of the $5S_{1/2}$, $6S_{1/2}$, $4D_{5/2}$ and $4D_{3/2}$ states in a lifetime of 3.45 μs. The ratio of the $5S_{1/2}$ decay channel to the other three channels is 1 to 4. When the Rb atoms decay to the $5S_{1/2}$ state, they emit fluorescence with a wavelength of 420 nm. Considering the lifetimes and the branching ratios, we find that emission of the blue fluorescence occurs at a rate of 2.9×10^5 s^{-1}. Observation of the blue fluorescence allows background-free detection of a small number of atoms.

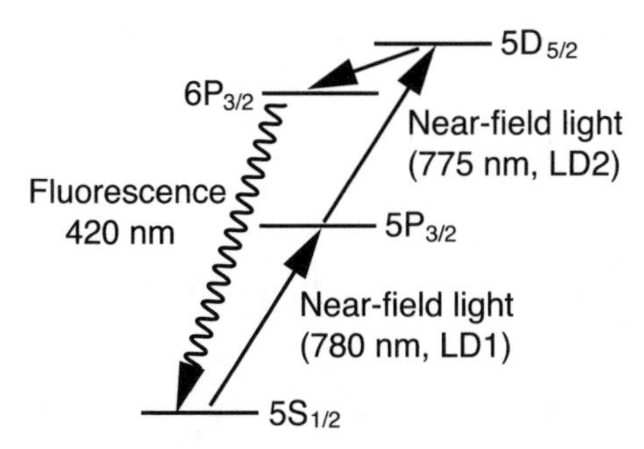

Fig. 19. Relevant energy levels of Rb. The first near-field light from a 780-nm diode laser (LD1) induces the $5S_{1/2} \rightarrow 5P_{3/2}$ transition. The second near-field light from a 775-nm diode laser (LD2) induces the $5P_{3/2} \rightarrow 5D_{5/2}$ transition. Some of the excited Rb atoms spontaneously decay to the $5S_{1/2}$ state via the $6P_{3/2}$ state and emit fluorescence with a wavelength of 420 nm

Experiment with Evanescent Light

In order to show the feasibility of the stepwise resonant excitation using two-color near-field lights, we conducted fluorescence spectroscopy with a Rb vapor cell. Figure 20 shows the experimental setup, where the Rb vapor cell is heated to 140°C. Two-color planar near-field lights (evanescent lights) are generated at a surface of a prism attached to a cell window by total-internal reflection of two counterpropagating diode-laser beams LD1 and LD2 at incident angles of 43° and 53°, respectively. Blue-fluorescence photons are collected by a fiber bundle with an efficiency of 0.02, and are guided to a photomultiplier tube (PMT) through a 420-nm interference filter (IF) with a bandwidth of 7 nm.

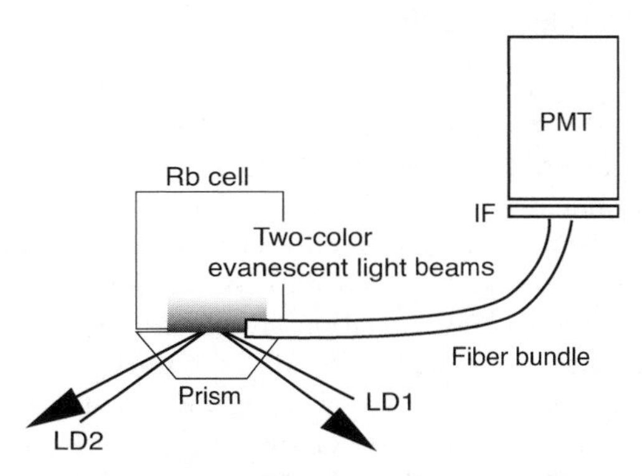

Fig. 20. Experimental setup with a Rb vapor cell. Two-color evanescent lights are induced at a surface of a prism attached to a cell window by total-internal reflection of two counterpropagating diode-laser beams LD1 and LD2. Fluorescence with a wavelength of 420 nm is collected by a fiber bundle and sent to a photomultiplier tube (PMT) with an interference filter (IF)

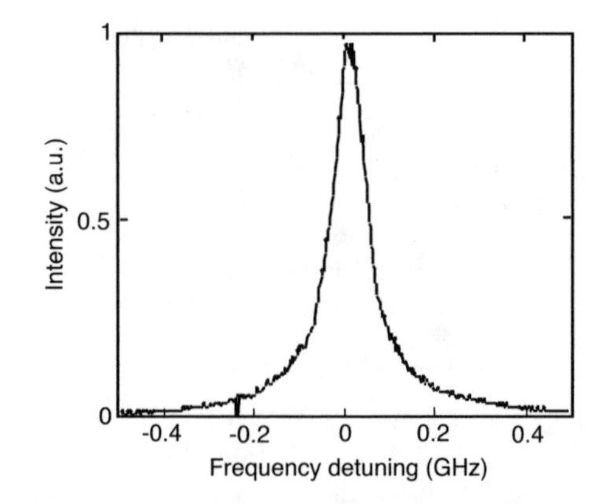

Fig. 21. Sub-Doppler spectrum of 420-nm fluorescence. The intensity is plotted as a function of the frequency detuning of LD2, where LD1 is locked to the $5\mathrm{S}_{1/2}$, $F = 3 \rightarrow 5\mathrm{P}_{3/2}$, $F = 4$ transition. The FWHM of the spectrum is 80 MHz

Figure 21 shows a fluorescence spectrum plotted as a function of the frequency detuning of LD2, where the frequency of LD1 is locked to the $5\mathrm{S}_{1/2}$, $F = 3 \rightarrow 5\mathrm{P}_{3/2}$, $F = 4$ hyperfine transition. The linewidth of the diode-laser beams is below 1 MHz and the beam diameter is 2 mm. The intensities of LD1 and LD2 are 2.9 $\mathrm{mW/cm^2}$ and 112 $\mathrm{mW/cm^2}$, respectively. The FWHM of the spectrum is 80 MHz. Note that the FWHM is narrower than the Doppler width of about 500 MHz. This is due to the configuration using two counterpropagating light beams. The sub-Doppler profile is determined by the natural linewidths of the relevant transitions and the transit-time broadening originating from the fact that atoms traverse the evanescent light.

Estimation of Detection Efficiency

Let us estimate the detection rate r_{bf} in the blue-fluorescence experiment mentioned above. If both Rabi frequencies of the $5\mathrm{S}_{1/2} \rightarrow 5\mathrm{P}_{3/2}$ and $5\mathrm{P}_{3/2} \rightarrow 5\mathrm{D}_{5/2}$ transitions are the same and much larger than both of the natural linewidths, the atomic decay from the $5\mathrm{P}_{3/2}$ state does not occur, so that the occupation probability of the $5\mathrm{D}_{5/2}$ state becomes maximum, i.e. 0.5. Note that the $5\mathrm{P}_{3/2}$ state is the so-called dark state. Since the density matrix elements with respect to the transitions from the $5\mathrm{S}_{1/2}$ and $5\mathrm{D}_{5/2}$ states are out of phase with each other, the $5\mathrm{P}_{3/2}$ state is not occupied due to the destructive interference [74]. In this case, the blue-fluorescence emission occurs most frequently. As the rate of the $5\mathrm{D}_{5/2} \rightarrow 6\mathrm{P}_{3/2} \rightarrow 5\mathrm{S}_{1/2}$ decay is $2.9 \times 10^5\ \mathrm{s}^{-1}$, it follows that the blue fluorescence is emitted at a rate of $1.5 \times 10^5\ \mathrm{s}^{-1}$. Then, from that, the collection efficiency of a lens with an NA exceeding 0.2 used here is 0.01 and the quantum efficiency of the PMT is 0.2 for 420 nm, we obtain $r_{\mathrm{bf}} = 3 \times 10^2\ \mathrm{s}^{-1}$.

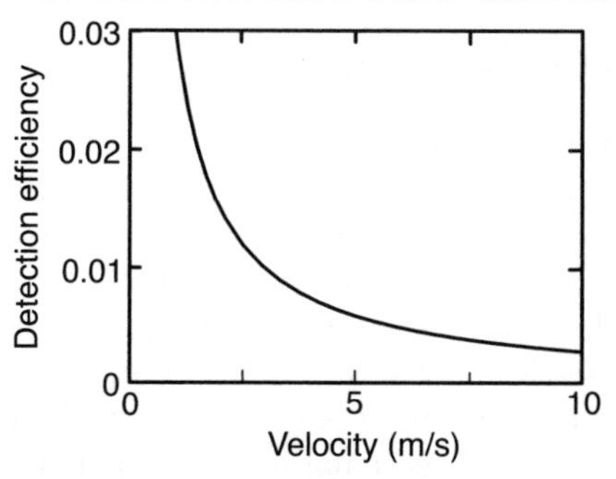

Fig. 22. Detection efficiency of Rb atoms plotted as a function of the atomic velocity

In the case where a Rb atom with a velocity of 1 m/s moves along the 100-μm long slit, the interaction time $t_{\rm int}$ with near-field light is 1×10^{-4} s. Consequently, the detection efficiency is approximately $r_{\rm bf} \times t_{\rm int} = 0.03$. The detection efficiency is proportional to the interaction time. Figure 22 shows the detection efficiency for slow Rb atoms. Although the detection efficiency is not necessarily high, it is sufficient for the deflection experiment using the slit-type deflector.

4 Guiding Cold Atoms through Hollow Light with Sisyphus Cooling

High-density slow atoms are required for manipulation using nanometric near-field light. Such a source can be generated as a cold atomic beam from MOT. Boosting by blue-detuned light was first reported in [75–78]. Then, guiding through hollow light [79–82] and funneling with a hollow prism [83,84] were reported. A hybrid method of funneling and guiding atoms by evanescent light has been also proposed [85,86]. In the hollow-light guiding, the atom flux intensity is obtained between 10^7 and 10^{11} atom/cm^2 s. By contrast, the mean velocity is between 2 and 50 m/s.

Generally, as the guide length becomes longer, the number of guided atoms decreases. The longest distance in the previous experiments was 18 cm and the guiding efficiency was 0.1 [87]. The guiding efficiency also depends on the mean temperature of atoms. Enhancement by PGC has been reported [88], in which the guiding efficiency was 0.6 over a distance of 14 cm. For longer guiding of atoms, Sisyphus cooling is effective [89]. In this case, atoms confined in a blue-detuned hollow light beam repeatedly lose the kinetic energy by transferring from the lower ground state to the upper ground state in reflection.

Here, we deal with atom guidance using a hollow light beam. The hollow-light guiding is a simple method of sending down cold atoms to the slit-type deflector. When Sisyphus cooling is performed, the guiding efficiency is 0.21 over a long distance of 26 cm.

4.1 Generation of Hollow Light

There are some methods of generating a hollow light beam. For example, a hologram [79,90,91], a double-cone prism [85], a hollow optical fiber [92,93], and interference of two Gaussian beams [94] have been used. A simple method is to use a filter with a Gaussian-distributed absorption character. We made a Gaussian filter with a dark spot of 2 mm by exposing a photographic plate to a TEM_{00}-mode He-Ne laser beam. The absorption coefficient α at the center is 0.97 for 780 nm and the power conversion efficiency is 0.6.

Figure 23 shows the cross-sectional images of a hollow light beam generated from a TEM_{00}-mode Ti:sapphire laser beam with a diameter of 10 mm. The hollow light beam was used in an experiment of guiding Rb atoms released from an MOT mentioned below. In the experiment, the hollow diameter at FWHM is reduced from 630 μm at the MOT (Fig. 23a) to 540 μm at 26 cm below the MOT (Fig. 23b) by two lenses with focal lengths of 1200 mm and 400 mm (see Fig. 26). The power of the hollow light beam is 500 mW.

The intensity distribution $I(r)$ of the generated hollow light beam is approximately given by

$$I(r) = \frac{I_0}{\pi\left[r_1^2 - \alpha r_1^2 r_2^2/(r_1^2 + r_2^2)\right]} \exp\left(-\frac{r^2}{r_1^2}\right)\left[1 - \alpha \exp\left(-\frac{r^2}{r_2^2}\right)\right] , \qquad (20)$$

where r is the distance from the beam axis, I_0 is the output intensity from the Gaussian filter, and r_1 and r_2 are the outer radius and the dark radius of the hollow light beam, respectively. From the diameters at FWHM, r_1 and r_2 are estimated to be 980 μm and 600 μm for Fig. 23a, and 830 μm and 500 μm for Fig. 23b. The hollow shape was able to be maintained over 60 cm.

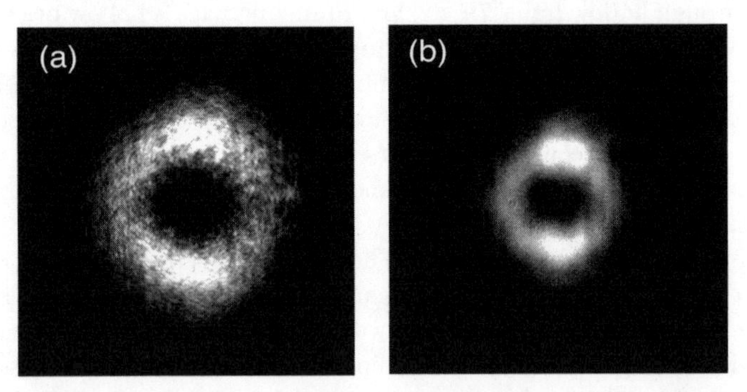

Fig. 23a,b. CCD-camera images of the cross section of a hollow light beam; (**a**) at an MOT and (**b**) at 26 cm below the MOT in the experiment of guiding Rb atoms

4.2 Sisyphus Cooling in Hollow Light

Principle

Under blue-detuning conditions, atoms are confined in a hollow light beam by a repulsive dipole force. If the atoms have two hyperfine ground states $|g1\rangle$ and $|g2\rangle$ and an excited state $|e\rangle$, which is called a Λ-type three-level system, one can perform Sisyphus cooling with the help of a pumping light [95–97].

Figure 24 shows the principle, where the dressed-atom picture [1] is used. For example, a ket $|g1, n\rangle$ expresses a state that an atom is in the lower ground state $g1$ and n photons exist. When the $g1$ atom enters light field, the bare state $|g1, n\rangle$ is changed to the dressed state $|1, n-1\rangle$ by the electric-dipole interaction. Three kets $|1, n\rangle$, $|2, n\rangle$, and $|3, n\rangle$ form a manifold. If the light frequency is blue-detuned by δ with respect to the lower ground state $g1$, the light-shift energies $\triangle E_1(r)$ for $g1$ and $\triangle E_2(r)$ for $g2$ are approximately given

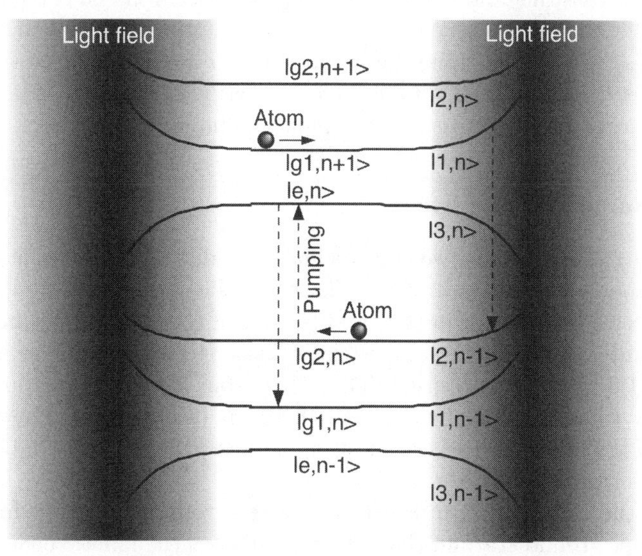

Fig. 24. Sisyphus cooling of a three-level atom with two ground states $g1$ and $g2$ and an excited state e in a hollow light. Three states $|g1, n+1\rangle$, $|g2, n+1\rangle$, and $|e, n\rangle$ with n photons in the dark region change to three dressed states $|1, n\rangle$, $|2, n\rangle$, and $|3, n\rangle$ in the bright region, respectively. Each atomic energy level is modulated by the light shift, which increases with the light intensity. The position-dependent light shift works as the repulsive potential for reflection. When the frequency detuning is positive for both ground states, the light shift for the upper ground state is smaller than that for the lower ground state. Consequently, if the transition from $|1, n\rangle$ to $|2, n-1\rangle$ occurs in the reflection, the atomic kinetic energy is lost. This process can be repeated by a pumping beam inducing the transition to the upper ground state

by [95]

$$\triangle E_1(r) \simeq \frac{2}{3}\frac{\hbar\Omega(r)^2}{4\delta} , \quad (21)$$

$$\triangle E_2(r) \simeq \frac{2}{3}\frac{\hbar\Omega(r)^2}{4(\delta+\delta_{\rm hfs})} , \quad (22)$$

where $\Omega(r)$ is the position-dependent Rabi frequency, and $\delta_{\rm hfs}$ is the hyperfine splitting between $g1$ and $g2$. The light shift equals the dipole-force potential given in (8) except a Clebsch–Gordan coefficient 2/3 for $\Omega(r)/\delta \ll 1$ because $\Omega(r)^2/\gamma^2 = I(r)/2I_{\rm s}$. As shown in Fig. 24, the light shift increases with the light intensity. If the maximum light shift is larger than the atomic kinetic energy, an atom is reflected after climbing up the potential slope and climbs down. Note that the light shift for the upper ground state $g2$ is smaller than that for the lower ground state $g1$. If an atom in the dressed state $|1,n\rangle$ transfers to the dressed state $|2,n-1\rangle$ by scattering a photon in the reflection, it loses the kinetic energy by the difference between two light shifts. Then, if the atom is returned from the upper ground state $g2$ to the lower ground state $g1$ by a pumping light tuned to the $|g2,n\rangle \rightarrow |e,n\rangle$ transition, the process that decreases atomic kinetic energy is repeated. This energy-reduction mechanism was named Sisyphus cooling after the Greek myth.

Estimation of Cooling Effects

Substituting (20) into (8), we can calculate the dipole-force potential generated by a hollow light beam. Figure 25 shows the change of the dipole-force potential for a ^{87}Rb atom in the cross section of the hollow light beam with $I_0 = 500$ mW shown in Fig. 23, where $\delta/2\pi = +1$ GHz, $\gamma/2\pi = 6.1$ MHz, and $I_{\rm s} = 1.6$ mW/cm^2 are used. The ^{87}Rb atom has two hyperfine ground states $F=1$ and 2 spaced by $\delta_{\rm hfs}/2\pi = 6.8$ GHz (see Fig. 27). The reflection potential $U_1(x)$ for the $F=1$ lower ground state is higher than $U_2(x)$ for the $F=2$ upper ground state.

When the ^{87} Rb atom in the $F=1$ lower ground state transfers to the $F=2$ upper ground state in the reflection, it loses the kinetic energy by the potential difference. For simplicity, we approximate the potentials shown in Fig. 25 using harmonic potentials as

$$U_1(x) \simeq \frac{1}{2}k_1x^2 , \quad U_2(x) \simeq \frac{1}{2}k_2x^2 , \quad (23)$$

where k_1 and k_2 are the spring constants. In this case, the mean energy $\langle\triangle U_{\rm s}\rangle$ that a ^{87}Rb atom loses in a Sisyphus-cooling event is given by

$$\langle\triangle U_{\rm s}\rangle = \frac{\int_0^{x_1}[U_1(x)-U_2(x)]\rho_{\rm e}(x)/\sqrt{x_1^2-x^2}{\rm d}x}{\int_0^{x_1}\rho_{\rm e}(x)/\sqrt{x_1^2-x^2}{\rm d}x} - \langle\triangle U_{\rm p}\rangle , \quad (24)$$

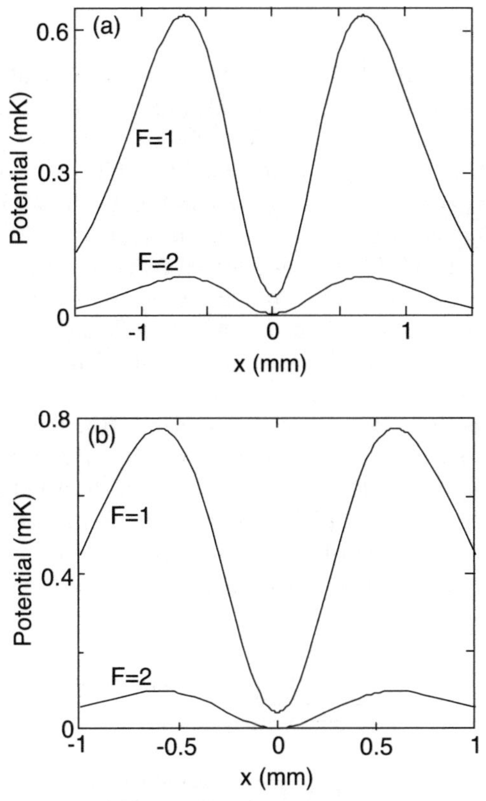

Fig. 25. Dipole-force potential in terms of temperature for ^{87}Rb in a cross section of a hollow light beam with a power of 500 mW; (**a**) for Fig. 23a, and (**b**) for Fig. 23b. The blue detuning is +1 GHz with respect to the $F = 1$ lower ground state

where $r_1 \simeq 500$ μm is the oscillation amplitude for $U_1(x)$, and $\rho_e(x) \sim k_1x^2/\hbar\delta$ is the population of the excited state. The mean energy $\langle\triangle U_p\rangle$ added to a ^{87}Rb atom in a pumping to the $F = 1$ lower ground state is given by

$$\langle\triangle U_p\rangle = \int_{-\infty}^{\infty} \sqrt{\frac{2}{\pi r_p^2}} \exp\left(-\frac{2x^2}{r_p^2}\right) [U_1(x) - U_2(x)]\mathrm{d}x \; , \qquad (25)$$

where $r_p = 150$ μm is the radius of the pumping light beam. From $k_1 = 1.9 \times 10^{-19}$ N/m and $k_2 = 2.4 \times 10^{-20}$ N/m, we obtain $\langle\triangle U_s\rangle \simeq 40$ μK in terms of temperature. Thus, the lateral motion of the guided atoms can be rapidly cooled.

4.3 Experiment

We conducted guiding of ^{87}Rb atoms using a hollow light beam shown in Fig. 23. Figure 26 shows the experimental configuration. A 5×10^6 ensemble is generated by an MOT under a background pressure of 9×10^{-10} torr and then cooled to a mean temperature of 42 μK by PGC. A hollow light beam is incident downward such that it does not illuminate the MOT. For Sisyphus cooling, a pumping light beam is also incident downward collinear to the hollow light beam. The ^{87}Rb atoms released from the MOT fall down while being guided by the hollow light beam, and they are detected at 26 cm below the MOT by two-step photoionization using a diode laser and an Ar-ion laser under a pressure of 1.3×10^{-10} torr. The ionized ^{87}Rb atoms are counted by a CEM applied at a negative bias of -3 kV.

Figure 27 shows the relevant energy levels of ^{87}Rb. The hollow light beam for guiding is blue-detuned by +1 GHz from the $5S_{1/2}$, $F = 1 \rightarrow 5P_{3/2}$, $F = 1$ transition. Note that the ^{87}Rb atoms released from the MOT are in the $F = 2$ upper ground state. By a 0.2-μW pumping light beam tuned to the $5S_{1/2}$, $F = 2 \rightarrow 5P_{3/2}$, $F = 2$ transition, the ^{87}Rb atoms are transferred to the $F = 1$ lower ground state. The ^{87}Rb atoms in the $F = 2$ upper ground state

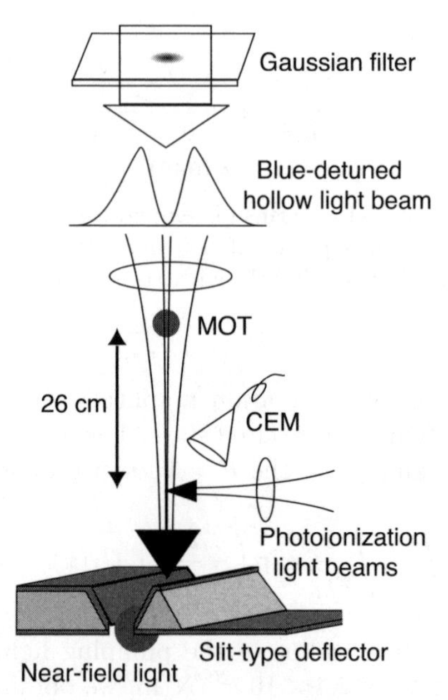

Fig. 26. Experimental configuration for guiding ^{87}Rb atoms. A blue-detuned hollow light beam formed through a Gaussian filter guides cold ^{87}Rb atoms generated by MOT in the vertical direction. A pumping light beam inducing Sisyphus cooling is also incident downward. The guided ^{87}Rb atoms are ionized at 26 cm below the MOT by photoionization light beams and detected by CEM. The cold atoms are sent to a slit-type deflector with near-field light

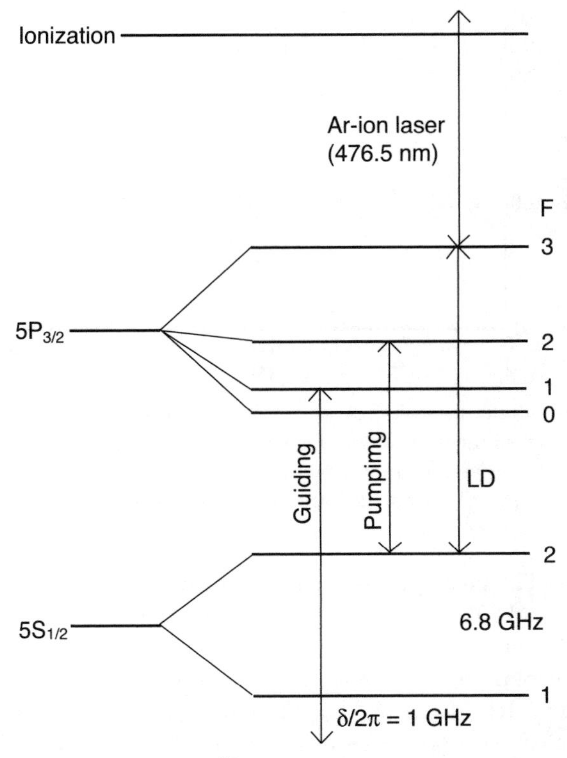

Fig. 27. Relevant energy levels of ^{87}Rb. A guiding light is blue-detuned by 1 GHz from the $5S_{1/2}$, $F = 1 \rightarrow 5P_{3/2}$, $F = 1$ transition. A pumping light is tuned to the $5S_{1/2}$, $F = 2 \rightarrow 5P_{3/2}$, $F = 2$ transition. Two-step photoionization is performed by using a diode laser (LD) tuned to the $5S_{1/2}$, $F = 2 \rightarrow 5P_{3/2}$, $F = 3$ transition and an Ar-ion laser with a wavelength of 476.5 nm. The two hyperfine ground states are seperated by 6.8 GHz

are selectively ionized by two lasers. First, a diode-laser beam with a power of 4.9 mW and a diameter of 220 μm excites the ^{87} Rb atoms in the $5S_{1/2}$, $F = 2$ upper ground state to the $5P_{3/2}$, $F = 3$ state. Then, a 476.5-nm Ar-ion laser beam with a power of 1 W and a diameter of 80 μm excites the ^{87} Rb atoms in the $5P_{3/2}$, $F = 3$ state to the ionization level at 4.18 eV above the $5S_{1/2}$ state.

Figure 28 shows the number of ^{87}Rb atoms in the $F = 2$ upper ground state plotted as a function of time t after release, where the time-of-flight signals are averaged over 10 measurements. Curve A shows the case of free fall. By contrast, curve B shows the case of guiding, and curve C shows the case of guiding with Sisyphus cooling. From this, the full width at the e^{-1} maximum in curve A is 23 ms, the initial mean temperature of the falling ^{87}Rb atoms is estimated to be 42 μK. The peak number of 50 and the total number of 3.0×10^3 in curve B are 2 and 7.5 times larger than those in curve

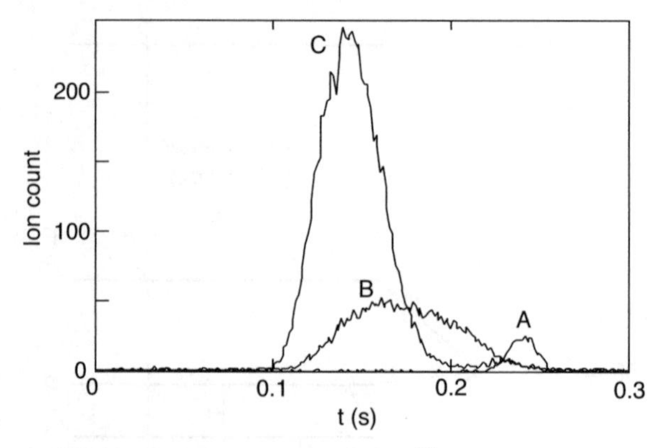

Fig. 28. Time-of-flight photoionization of the ^{87}Rb atoms in the $F = 2$ upper ground state, where ion count averaged over 10 measurements is plotted as a function of time t after release. *Curve A* shows the case of free fall. *Curve B* shows the case of guiding. *Curve C* shows the case of guiding with Sisyphus cooling

A, respectively. The peak number of 240 and the total number of 8.2×10^3 in curve C are 9.6 and 20 times larger than those in curve A, respectively, and 4.8 and 2.7 times larger than those in curve B, respectively. These results indicate that Sisyphus cooling contributes enhancement of the guiding efficiency. Note that, when the ^{87}Rb atoms are transferred to the lower ground state, the confinement effect also grows because the repulsive potential for the lower ground state is higher than that for the upper ground state. The increase of the guided atoms is the result of cooperation of Sisyphus cooling and tight confinement.

The peak in curve B appears earlier than that in curve A. This is due to scattering of photons of the guide light beam. Since the guide light beam propagates downward, it accelerates the falling atoms. In curve C, atoms are more accelerated by scattering of photons of an additional pumping light beam. In order to avoid the heating effects, the guiding and pumping light beams should be directed upwards. In this case, cooling effects are expected. The experiments are currently in progress.

4.4 Estimation of Atom Flux

Figure 29 shows the spatial distribution of the guided ^{87}Rb atoms obtained by scanning the cross section with photoionization light beams in the case with Sisyphus cooling. If the guided atoms have a Gaussian distribution, the number of atoms $\Phi(x_\mathrm{i})$ obtained when the axis of the photoionization light beams is at x_i is given by

$$\Phi(x_\mathrm{i}) = \frac{N_\mathrm{a}}{\pi r_\mathrm{a}^2} \int_{x_\mathrm{i}-r_\mathrm{i}}^{x_\mathrm{i}+r_\mathrm{i}} \exp\left(-\frac{x^2}{r_\mathrm{a}^2}\right) \mathrm{d}x \int_{-\infty}^{\infty} \exp\left(-\frac{y^2}{r_\mathrm{a}^2}\right) \mathrm{d}y \,, \qquad (26)$$

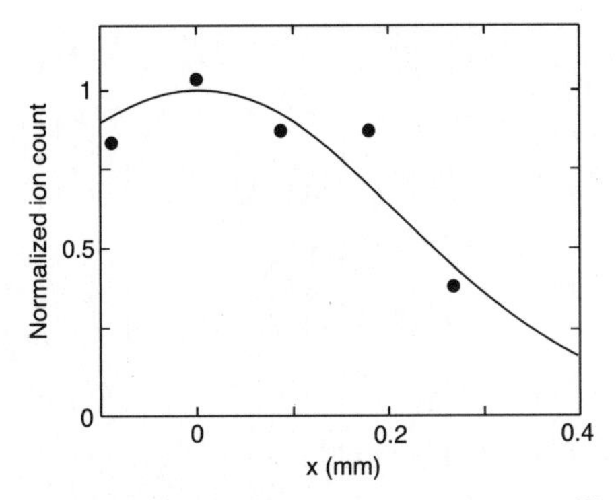

Fig. 29. Cross-sectional spatial distribution of the guided ^{87}Rb atoms at 26 cm below MOT. *Full circles* indicate the experimental values, while the *solid curve* is a fitting. The ion count is normalized to the value at the center ($x = 0$) of the hollow light beam

where N_a, r_i, and r_a are the number of atoms trapped at MOT, the radius of the photoionization light beams, and the distribution radius of the guided atoms, respectively. The x axis denotes the scanning direction in the horizontal cross section and the y axis is taken to be perpendicular to the x axis in the cross section. The origin $x = 0$ is at the center of the hollow light beam. The solid curve in Fig. 29 shows the fitting for $N_a = 5 \times 10^6$, $r_i = 40$ μm, and $r_a = 300$ μm. The FWHM is 550 μm, which closely matches the dark spot 540 μm of the hollow light beam.

The guiding efficiency is given by

$$\eta_g = \frac{N_i}{\eta_e q \Phi(0)} , \tag{27}$$

where N_i, η_e, and q are the number of the ionized atoms for $x_i = 0$, the photoionization efficiency, and the quantum efficiency of the CEM. Using $N_i = 8.2 \times 10^3$, $\eta_e = 5.7 \times 10^{-2}$, and $q = 0.9$, we obtain $\eta_g = 0.21$. We successively generated cold atoms every 23 ms. In this case, the mean flux intensity of the cold ^{87}Rb atoms is estimated to be 1.9×10^{10} atom/cm^2 s. This value is sufficient for the atom-deflection experiments.

5 Outlook

Atom optics using near-field light has many applications including nanofabrication and mesoscopic science. We can make several unique devices for atom manipulation. For example, we are making an array composed of nanometric slits. Figure 30 shows a slit array forming a matrix of 300 × 40, which was fabricated on a substrate with a size of 15 μm × 115 μm. The typical slit width is 200 nm, and each slit has a V-shaped groove as shown in Fig. 30b.

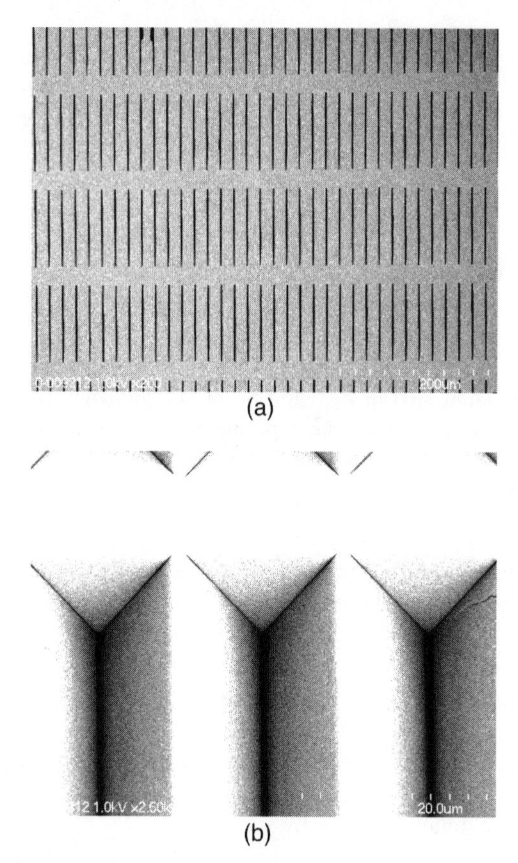

Fig. 30. (**a**) SEM image of a slit-array-type atom detector composed of a matrix of 300×40, where the typical slit width is 200 nm. The size of the substrate is 15 μm × 115 μm. (**b**) SEM image of the backside, where each slit has a V-shaped groove

The slit array is used as an atom detector with higher performance than the monoslit-type atom detector. Indeed, using two-step photoionization, the slit-array-type atom detector can work at a detection rate of about 10 atom/s for the cold atomic beam mentioned in Sect. 4. The slit array can be also used as an atom grating.

The slit-type atom deflector presented in Sect. 2 was developed for atom deposition. Since a cold atomic beam has a narrow velocity distribution, as shown in Fig. 28, we can move the cold atoms to the position aimed at with high accuracy using the deflector. However, a cold atom has a wave character, as quantum mechnics shows, so that the spatial accuracy in the atom manipulation is limited by the diffraction limit of the de Broglie wave, which is about 10 nm. In order to avoid the uncertainty of the de Broglie wave, we should generate a cold-atom ensemble just in front of a substrate for atom deposition [98]. To this end, we are developing a nanometric atom funnel using near-field light. The nanofunnel is fabricated from a pyramidal probe. The size of the exit hole is comparable to the atomic de Broglie wavelength and only an atom is selectively outputted. By setting the funnel just above a substrate, one can efficiently deposit atoms at the exact position aimed at.

We are also developing an atom trap using a fiber probe [52,53]. The atom trap uses a balance between a repulsive dipole force and an attractive van der Waals force near the nanometric aperture of the fiber probe. The single-atom trapping technique can be used for atom-by-atom deposition. By operating the fiber probe, one can move an atom to a position on a substrate. The atom is pushed toward the position by increasing the light intensity.

Interaction between atoms and near-field light has not been fully examined. The atom-manipulation techniques work for the study of the interaction in detail. Note that near-field light is usually observed by transformation into far-field light. Consequently, a fiber probe measures near-field light by destroying it. By contrast, atoms can see near-field light without destroying it; for example, the original intensity distribution of near field light is measurable by examining the scattering of atoms in the reflection scheme. The atom-control techniques presented here will greatly develop nanophotonics and create a new research field, atom-photonics.

References

1. C. Cohen-Tannoudji, J. Dupont-Roc, G. Grynberg: *Atom-Photon Interaction* (John Wiley and Sons, New York 1992)
2. 'The Mechanical Effects of Light'. In: Special issue ed. by P. Meystre, S. Stenholm of J. Opt. Soc. Am. B **2** (1985) pp. 1707–1860
3. 'Laser Cooling and Trapping of Atoms'. In: Special issue ed. by S. Chu, C. Wieman of J. Opt. Soc. Am. B **6** (1989) pp. 2023–2278
4. *Laser Manipulation of Atoms and Ions*, ed. by E. Arimondo, W.D. Phillips, F. Strumia (North-Holland, Amsterdam 1992)
5. V.G. Minogin, V.S. Letokhov: *Laser Light Pressure on Atoms* (Gordon and Breach Science Publishers, New York 1987)
6. A.P. Kazantsev, G.I. Surdutovich, V.P. Yakovlev: *Mechanical Action of Light on Atoms* (World Scientific, Singapore 1990)
7. E.L. Raab, M. Prentiss, A. Cable, S. Chu, D.E. Pritchard: Phys. Rev. Lett. **59**, 2631 (1987)
8. C. Monroe, W. Swann, H. Robinson, C. Wieman: Phys. Rev. Lett. **65**, 1571 (1990)
9. C.N. Cohen-Tannoudji, W.D. Phillips: Phys. Today **43**, No. 10, 33 (1990)
10. H.J. Metcalf, P. van der Straten: *Laser Cooling and Trapping* (Springer, New York 1999)
11. J.W.R. Tabosa, G. Chen, Z. Hu, R.B. Lee, H.J. Kimble: Phys. Rev. Lett. **66**, 3245 (1991)
12. P.S. Jessen, I.H. Deutsch: 'Optical Lattices'. In: *Advances in Atomic, Molecular, and Optical Physics* **37**, ed. by B. Bederson, H. Walther (Academic Press, San Diego 1996) pp. 95–138

13. T. Walker, P. Feng: 'Measurements of Collisions between Laser-Cooled Atoms'. In: *Advances in Atomic, Molecular, and Optical Physics* **34**, ed. by B. Bederson, H. Walther (Academic Press, San Diego 1994) pp. 125–170
14. P.S. Julienne, A.M. Smith, K. Burnett: 'Theory of Collisions between Laser Cooled Atoms'. In: *Advances in Atomic, Molecular, and Optical Physics* **30**, ed. by D.B. Bates, B. Bederson (Academic Press, San Diego 1992) pp. 141–198
15. J. Weiner: 'Advances in Ultracold Collisions: Experimentation and Theory'. In: *Advances in Atomic, Molecular, and Optical Physics* **35**, ed. by B. Bederson, H. Walther (Academic Press, San Diego 1995) pp. 45–78
16. M.A. Kasevich, E. Riis, S. Chu, R.G. Devoe: Phys. Rev. Lett. **63**, 612 (1989)
17. *Atom Interferometry*, ed. by P.R. Berman (Academic Press, San Diego 1997)
18. C.S. Adams, O. Carnal, J. Mlynek: 'Atom Interferometry'. In: *Advances in Atomic, Molecular, and Optical Physics* **34**, ed. by B. Bederson, H. Walther (Academic Press, San Diego 1994) pp. 1–34
19. J. Fujita, M. Morinaga, T. Kishimoto, M. Yasuda, S. Matsui, F. Shimizu: Nature **380**, 691 (1996)
20. F. Shimizu: 'Atom Holography'. In: *Advances in Atomic, Molecular, and Optical Physics* **42**, ed. by B. Bederson, H. Walther (Academic Press, San Diego 2000) pp. 182–260
21. M.Z. Raizen: 'Quantum Chaos with Cold Atoms'. In: *Advances in Atomic, Molecular, and Optical Physics* **41**, ed. by B. Bederson, H. Walther (Academic Press, San Diego 1999) pp. 43–82
22. M.R. Anderson, J.R. Ensher, M.R. Matthews, C.E. Wieman, E.A. Cornell: Science **269**, 198 (1995)
23. *Bose–Einstein Condensation in Atomic Gases*, ed. by M. Inguscio, S. Stringari, C.E. Wieman (IOS Press, Amsterdam 1999)
24. C.J. Pethick, H. Smith: *Bose–Einstein Condensation in Dilute Gases* (Cambridge University Press, Cambridge 2002)
25. M.R. Andrews, C.G. Townsend, H.-J. Miesner, D.S. Durfee, D.M. Kurn, W. Ketterle: Science **275**, 637 (1997)
26. C. Cohen-Tannoudji: 'Atomic motion in laser light'. In: *Fundamental Systems in Quantum Optics*, ed. by J. Dalibard, J.M. Raimond, J. Zinn-Justin (Elsevier, Amsterdam 1992) pp. 1–164
27. V.I. Balykin, V.S. Letokhov: *Atom Optics with Laser Light* (Harwood Academic, Chur 1995) Chap. 4
28. P. Meystre: *Atom Optics* (Springer, New York 2001)
29. C.M. Savage, S. Marksteiner, P. Zoller: 'Atomic Waveguides and Cavities from Hollow Optical Fibers'. In: *Fundamentals of Quantum Optics III*, ed. by F. Ehlotzky (Springer, Berlin 1993) pp. 60–74
30. S. Marksteiner, C.M. Savage, P. Zoller, S.L. Rolston: Phys. Rev. A **50**, 2680 (1994)
31. H. Ito, K. Sakaki, T. Nakata, W. Jhe, M. Ohtsu: Opt. Commun. **115**, 57 (1995)
32. H. Ito, K. Sakaki, T. Nakata, W. Jhe, M. Ohtsu: Ultramicroscopy **61**, 91 (1995)
33. M.J. Renn, E.A. Donley, E.A. Cornell, C.E. Wieman, D.Z. Anderson: Phys. Rev. A **53**, R648 (1996)
34. H. Ito, T. Nakata, K. Sakaki, W. Jhe, M. Ohtsu: Phys. Rev. Lett. **76**, 4500 (1996)
35. H. Ito, K. Sakaki, W. Jhe, M. Ohtsu: Opt. Commun. **141**, 43 (1997)

36. J.P. Dowling, J. Gee-Banacloche: 'Evanescent Light-Wave Atom Mirrors, Resonators, Waveguides, and Traps'. In: *Advances in Atomic, Molecular, and Optical Physics* **37**, ed. by B. Bederson, H. Walther (Academic Press, San Diego 1996) pp. 1–94
37. V.I. Balykin: 'Atom Waveguide'. In: *Advances in Atomic, Molecular, and Optical Physics* **41**, ed. by B. Bederson, H. Walther (Academic Press, San Diego 1999) pp. 182–260
38. K. Sengstock, W. Ertmer: 'Laser Manipulation of Atoms'. In: *Advances in Atomic, Molecular, and Optical Physics* **35**, ed. by B. Bederson, H. Walther (Academic Press, San Diego 1995) pp. 1–34
39. G. Timp, R.E. Behringer, D.M. Tennant, J.E. Cunningham, M. Prentiss, K.K. Berggren: Phys. Rev. Lett. **69**, 1636 (1992)
40. M. Prentiss, G. Timp, N. Bigelow, R.E. Behringer, J.E. Cunningham: Appl. Phys. Lett. **60**, 1027 (1992)
41. J.J. McClelland, R.E. Scholten, E.C. Palm, R.J. Celotta: Science **262**, 877 (1993)
42. R. Gupta, J.J. McClelland, Z.J. Jabbour, R.J. Celotta: Appl. Phys. Lett. **67**, 1378 (1995)
43. R.W. McGowan, D.M. Giltner, S.A. Lee: Opt. Lett. **20**, 2535 (1995)
44. W.R. Anderson, C.C. Bradley, J.J. McClelland, R.J. Celotta: Phys. Rev. A **59**, 2476 (1999)
45. Th. Schulze, T. Müther, D. Jürgens, B. Brezger, M.K. Oberthaler, T. Pfau, J. Mlynek: Appl. Phys. Lett. **78**, 1781 (2001)
46. E. Jurdik, J. Hohlfeld, H. van Kempen, Th. Rasing, J.J. McClelland: Appl. Phys. Lett. **80**, 4443 (2002)
47. K.K. Berggren, A. Bard, J.L. Wilbur, J.D. Gillaspy, A.G. Helg, J.J. McClelland, S.L. Rolston, W.D. Phillips, M. Prentiss, G.M. Whitesides: Science **269**, 1255 (1995)
48. K.S. Johnson, K.K. Berggren, A. Black, C.T. Black, A.P. Chu, N.H. Dekker, D.C. Raiph, J.H. Thywissen, R. Younkin, M. Tinkham, M. Prentiss, G.M. Whitesides: Appl. Phys. Lett. **69**, 2773 (1996)
49. M. Ohtsu: J. Lightwave Tecnol. **13**, 1200 (1995)
50. M. Ohtsu, H. Hori: *Near-Field Nano-Optics* (Kluwer/Plenum, New York 1999) Chap 8
51. K. Kobayashi, M. Ohtsu: J. Microsc. **194**, 249 (1999)
52. H. Ito, M. Ohtsu: 'Near-Field Optical Atom Manipulation: Toward Atom Photonics'. In: *Near-Field Nano/Atom Optics and Technology*, ed. by M. Ohtsu (Springer, Tokyo 1998) Chap. 11
53. M. Ohtsu, K. Kobayashi, H. Ito, G.-H. Lee: Proc. of the IEEE **88**, 1499 (2000)
54. K. Kobayashi, S. Sangu, H. Ito, M. Ohtsu: 'Effective probe-sample onteraction: Toward atom deflection and manipulation'. In: *Near-Field Optics: Principles and Applications, 2nd Asia-Pacific Workshop on Near Field Optics, 2000*, ed. by X. Zhu, M. Ohtsu, 82 (World Scientific, Singapore 2000) pp. 82–88
55. K. Kobayashi, S. Sangu, H. Ito, M. Ohtsu: Phys. Rev. A **63**, 013806 (2001)
56. K. Totsuka, H. Ito, K. Suzuki, K. Yamamoto, M. Ohtsu, T. Yatsui: Appl. Phys. Lett. **82**, 1616 (2003)
57. K. Totsuka, H. Ito, T. Kawamura, M. Ohtsu: Jpn. J. Appl. Phys. **41**, 1566 (2002)
58. H. Ito, K. Sakaki, M. Ohtsu, W. Jhe: Appl. Phys. Lett. **70**, 2496 (1997)

59. T. Yatsui, M. Kourogi, K. Tsutsui, M. Ohtsu, J. Takahashi: Opt. Lett. **25**, 1279 (2000)
60. P.N. Minh, T. Ono, M. Esashi: Rev. Sci. Instrum. **71**, 3111 (2000)
61. H.U. Danzebrink, Th. Dziomba, T. Sulzbach, O. Ohlsson, C. Lehrer, L. Frey: J. Microsc. **194**, 335 (1999)
62. R.G. Newton: *Scattering Theory of Waves and Particles*, 2nd edn. (Springer, New York 1982)
63. M. Chevrollier, M. Fichet, M. Oria, G. Rahmat, D. Bloch, M. Ducloy: J. Phys. II France **2**, 631 (1992)
64. W. Jhe, J.W. Kim: Phys. Rev. A **51**, 1150 (1995)
65. H. Nha, W. Jhe: Phys. Rev. A **54**, 3505 (1996)
66. Ch. Kurtsiefer, J. Mlynek: Appl. Phys. B **64**, 85 (1997)
67. M.D. Scheer, J. Fine: J. Chem. Phys. **39**, 1752 (1963)
68. S. Datz, E.H. Taylor: J. Chem. Phys. **25**, 389 (1956)
69. K. Totsuka, H. Ito, M. Ohtsu: IEICE Trans. Electron. **E85-C**, 2093 (2002)
70. V.S. Letokhov: *Laser Photoionization Spectroscopy* (Academic Press, San Diego 1987)
71. S. Feron, J. Reinhardt, M. Ducloy, O. Gorceix, S. Nic Chormaic, Ch. Miniatura, J. Robert, J. Baudon, V. Lorent, H. Haberland: Phys. Rev. A **49**, 4733 (1994)
72. F. de Fornel: *Evanescent Waves* (Springer, Berlin 2001) p. 9
73. R. Kaiser, Y. Levy, N. Vansteenkiste, A. Aspect, W. Seifert, D. Leipold, J. Mlynek: Opt. Commun. **104**, 234 (1994)
74. M.O. Scully, M.S. Zubairy: *Quantum Optics* (Cambridge University Press, Cambridge 1997) Chap. 7
75. Z.T. Lu, K.L. Corwin, M.J. Renn, M.H. Anderson, E.A. Cornell, C.E. Wieman: Phys. Rev. Lett. **77**, 3331 (1996)
76. S. Weyers, E. Aucouturire, C. Valentin, N. Dimarcq: Opt. Commun. **143**, 30 (1997)
77. R.J.C. Spreeuw, M. Weidemüller, J.T.M. Walraven: Phys. Rev. A **58**, 3891 (1998)
78. J. Schoser, A. Batär, R. Löw, V. Schweikhard, A. Grabowski, Yu.B. Ovchinikov, T. Pfau: Phys. Rev. A **66**, 023410 (2002)
79. T. Kuga, Y. Torii, N. Shiokawa, T. Hirano: Phys. Rev. Lett. **78**, 4713 (1997)
80. X. Xu, V. G. Minogin, K. Lee, W. Jhe: Phys. Rev. A **60**, 4796 (1999)
81. X. Xu, K. Kim, W. Jhe, N. Kwan: Phys. Rev. A **63**, 063401 (2001)
82. M. Yan, Y. Yin, Y. Zhu: J. Opt. Soc. Am. B **17**, 1817 (2000)
83. K.H. Kim, K.I. Lee, H.R. Noh, W. Jhe, N. Kwon, M. Ohtsu: Phys. Rev. A **64**, 013402 (2001)
84. A. Camposeo, A. Piombini, F. Cervelli, F. Tantussi, F. Fuso, E. Arimondo: Opt. Commun. **200**, 231 (2001)
85. H. Ito, K. Sakaki, W. Jhe, M. Ohtsu: Phys. Rev. A **56**, 712 (1997)
86. A. Takamizawa, H. Ito, M. Ohtsu: Jpn. J. Appl. Phys. **41**, 6215 (2002)
87. Y. Song, D. Milan, T. Hill III: Opt. Lett. **15**, 1805 (1999)
88. S. Kuppens, M. Rauner, M. Schiffer, K. Sengstock, W. Ertmer: Phys. Rev. A **58** (1998)
89. J. Yin, Y. Zhu: J. Opt. Soc. Am. B **15**, 2235 (1998)
90. H.S. Lee, B.W. Stewart, K. Choi, H. Fenichel: Phys. Rev. A **49**, 4922 (1994)
91. K.T. Gahagan, G.A. Swartzlander, Jr: Opt. Commun. **21**, 827 (1996)
92. J. Yin, H.-R. Noh, K.-I. Lee, K.-H. Kim, Y.-Z. Wang, W. Jhe: Opt. Commun. **138**, 287 (1997)

93. A. Takamizawa, H. Ito, M. Ohtsu: Jpn. J. Appl. Phys. **39**, 6737 (2000)
94. S.M. Iftiquar, H, Ito, M Ohtsu: 'Tunable Doughnut Light Beam for a Near-Field Optical Funnel of Atoms'. In: *4th Pacific Rim Conference on Lasers and Electro-Optics 2001, Postdeadline Paper* (IEEE, Chiba 2001) pp. 34–35
95. J. Söding, R. Grimn, Yu.B. Ovchinnikov: Opt. Commun. **119**, 652 (1995)
96. P. Desbiolles, M. Arndt, P. Szriftgiser, J. Dalibard: Phys. Rev. A **54**, 4292 (1996)
97. H. Nha, W. Jhe: Phys. Rev. A **56**, 729 (1997)
98. J. Reichel, W. Hänsel, T.W. Hänsch: Phys. Rev. Lett. **83**, 3398 (1999)

Index

Springer Series in
OPTICAL SCIENCES

New editions of volumes prior to volume 70

1 **Solid-State Laser Engineering**
By W. Koechner, 5th revised and updated ed. 1999, 472 figs., 55 tabs., XII, 746 pages

14 **Laser Crystals**
Their Physics and Properties
By A. A. Kaminskii, 2nd ed. 1990, 89 figs., 56 tabs., XVI, 456 pages

15 **X-Ray Spectroscopy**
An Introduction
By B. K. Agarwal, 2nd ed. 1991, 239 figs., XV, 419 pages

36 **Transmission Electron Microscopy**
Physics of Image Formation and Microanalysis
By L. Reimer, 4th ed. 1997, 273 figs. XVI, 584 pages

45 **Scanning Electron Microscopy**
Physics of Image Formation and Microanalysis
By L. Reimer, 2nd completely revised and updated ed. 1998,
260 figs., XIV, 527 pages

Published titles since volume 70

70 **Electron Holography**
By A. Tonomura, 2nd, enlarged ed. 1999, 127 figs., XII, 162 pages

71 **Energy-Filtering Transmission Electron Microscopy**
By L. Reimer (Ed.), 1995, 199 figs., XIV, 424 pages

72 **Nonlinear Optical Effects and Materials**
By P. Günter (Ed.), 2000, 174 figs., 43 tabs., XIV, 540 pages

73 **Evanescent Waves**
From Newtonian Optics to Atomic Optics
By F. de Fornel, 2001, 277 figs., XVIII, 268 pages

74 **International Trends in Optics and Photonics**
ICO IV
By T. Asakura (Ed.), 1999, 190 figs., 14 tabs., XX, 426 pages

75 **Advanced Optical Imaging Theory**
By M. Gu, 2000, 93 figs., XII, 214 pages

76 **Holographic Data Storage**
By H.J. Coufal, D. Psaltis, G.T. Sincerbox (Eds.), 2000
228 figs., 64 in color, 12 tabs., XXVI, 486 pages

77 **Solid-State Lasers for Materials Processing**
Fundamental Relations and Technical Realizations
By R. Iffländer, 2001, 230 figs., 73 tabs., XVIII, 350 pages

78 **Holography**
The First 50 Years
By J.-M. Fournier (Ed.), 2001, 266 figs., XII, 460 pages

79 **Mathematical Methods of Quantum Optics**
By R.R. Puri, 2001, 13 figs., XIV, 285 pages

Springer Series in

OPTICAL SCIENCES

80 **Optical Properties of Photonic Crystals**
By K. Sakoda, 2001, 95 figs., 28 tabs., XII, 223 pages

81 **Photonic Analog-to-Digital Conversion**
By B.L. Shoop, 2001, 259 figs., 11 tabs., XIV, 330 pages

82 **Spatial Solitons**
By S. Trillo, W.E. Torruellas (Eds), 2001, 194 figs., 7 tabs., XX, 454 pages

83 **Nonimaging Fresnel Lenses**
Design and Performance of Solar Concentrators
By R. Leutz, A. Suzuki, 2001, 139 figs., 44 tabs., XII, 272 pages

84 **Nano-Optics**
By S. Kawata, M. Ohtsu, M. Irie (Eds.), 2002, 258 figs., 2 tabs., XVI, 321 pages

85 **Sensing with Terahertz Radiation**
By D. Mittleman (Ed.), 2003, 207 figs., 14 tabs., XVI, 337 pages

86 **Progress in Nano-Electro-Optics I**
Basics and Theory of Near-Field Optics
By M. Ohtsu (Ed.), 2003, 118 figs., XIV, 161 pages

87 **Optical Imaging and Microscopy**
Techniques and Advanced Systems
By P. Török, F.-J. Kao (Eds.), 2003, 260 figs., XVII, 395 pages

88 **Optical Interference Coatings**
By N. Kaiser, H.K. Pulker (Eds.), 2003, 203 figs., 50 tabs., XVI, 504 pages

89 **Progress in Nano-Electro-Optics II**
Novel Devices and Atom Manipulation
By M. Ohtsu (Ed.), 2003, 115 figs., XIII, 188 pages